Lean Enterprise Inventory Strategies

ASWO!

("__A__h __S__hucks, __W__e're __O__ut!")

Eric Matson
Principal Engineer (Retired)
Global Lean Operations Group, Fortune-50 Corp.

Sr. Partner and Co-Founder
Integrated Lean Systems Deployment, llc® (ILSD)

AuthorHouse™
1663 Liberty Drive
Bloomington, IN 47403
www.authorhouse.com
Phone: 1-800-839-8640

First published by AuthorHouse 9/9/2009

ISBN: 978-1-4490-1121-5 (sc)
ISBN: 978-1-4490-1122-2 (e)

Library of Congress Control Number: 2009909286

*About the cover: Marketing is fond of the adage "You can't sell from an empty wagon!", but a full one is very
costly. "ASWO!" explores balancing strategies that support lean value streams.*

*Additional copies of "ASWO!" may be ordered from Amazon.com, BarnesAndNoble.com, or directly from
Authorhouse.com.*

Printed in the United States of America
Bloomington, Indiana

This book is printed on acid-free paper.

To "Dutch"

(Who thinks I watch "Judge Judy" on TV all day)

Also, personal and profound thanks to Dan Faurie, my friend, my partner, my Editor, and my Brother. Merci bien, mon ami!

About the Author

At the end of 2008, Eric retired from 40+ years at UTC/Carrier Corp., where he had served as the Principal Engineer of their Global Lean Operations Group since 1999. In this role, he worked closely with most of the company's nearly 100 factories and many of its suppliers and 160+ wholesale distribution centers around the World. As a UTC-certified "***Master-Level*** Operations Transformation Leader", he served in the "Lean Sensei" role for all aspects of Lean Enterprise development, including Product Design, Manufacturing and Wholesale Distribution, and he acquired an informal specialist reputation in the area of Materials and Inventory Management.

He has been an international conference Speaker for both **A.P.I.C.S.** (The Association for Operations Management) and the (now integrated) **I.B.F.** (Institute of Business Forecasting), and has authored numerous technical articles for European and Far-Eastern technical journals.

Eric graduated with a B.S.I.E. from Lehigh University in 1968, and did Graduate Business studies at Syracuse University before accepting (3) overseas living/working assignments in Europe and the Far East during the1970-1987 period.

Prior to that time, his varied assignments provided him with in-depth experiences of Manufacturing, Product Application, Product Design, Marketing & Sales, Field Service, Computer-Aided Engineering, and Technical Training. From 1987-96, he also worked in the arenas of Corp. HR / Operations Development and Corporate Quality (including a stint as the Corporation Manager of Customer Satisfaction).

Since "re-tiring" (hyphenated, as in "tired again"?), Eric has been busy as Co-Founder of "Integrated **L**ean **S**ystems **D**eployment (**ILSD** llc)", a consulting partnership focused on bringing "Lean" methods and product designs to smaller commercial operations, which are often on quite limited advisor budgets (see www.ILSDllc.com).

Eric has been happily married (to the same great Lady) since 1968, is the father of two, and the Grandfather of seven (and still counting?). His hobbies include reading, clay target shooting, radio-controlled sail boat racing, and writing guides such as this one.

Introduction

Rule 1: **NEVER** stop production or block a sale for lack of materials and   product.
Rule 2: **ALWAYS** work to minimize material and product inventories.
Rule 3: **AVOID** potentially dysfunctional or counterproductive performance metrics.
Rule 4: **LEARN** from every situation and player (including the occasional idiot).
Rule 5: ~~IF~~ **When** **CONFUSED or FRUSTRATED**, you probably ignored Rule 3.

If you are reading this guide, you are probably directly responsible for inventory, or are responsible for the folks who are. You may be coping with seemingly chronic shortages, even though your warehouses and storage areas are overflowing with (apparently unneeded?) "something". Sales are tight, but the call for profit growth persists, so costs need to get lower, and inventory is a prime but intimidating candidate. Perhaps you've been "Leaning" the other aspects of operations, but aren't sure how to shrink inventories without disrupting production. Relax… Things could be worse (they could have stuck me with the job)!

For every apparent overstock identified, there seems to be a long, sad story explaining why it is there… some terrific quantity-buy… the supplier is stopping production… hedging future price increases… "global sourcing" requires "container loads"… outsourcing buffer needs… uncertain **F**orecasts (the F-Word?)… unreliable quality, deliveries and lead times… etc., etc., but the end result is a **BIG**, long sad story.

If I had all the right answers, I'd be independently wealthy by now. Unfortunately, nobody seems to be that bright, and the "right answers" can prove situational. I have, however, acquired a long and thought-provoking set of "interesting questions", helpful tools, and insights to mistakes & successes by others, which I propose to share here. The ideas for leaning-out production inventories are pretty universal, but I will add and identify any additional thoughts related to finished goods and/or replacement components as appropriate.

I'm not a fan of political or corporate correctness, and notoriously "call 'em as I see 'em". I don't especially care *who* made the decisions behind any current messes (the conditions were different back then)… but I DO care that we understand *what* went wrong and put it right. In my former role with a major, global corporation, I was sometimes called "NITO-san", but more often referred to as "Doctor Duck"… i.e., "If it is the size of a soccer ball, is covered with feathers, waddles when it walks, and quacks, it is a *DUCK*… *NOT* a baby swan or fledgling peacock!" Accordingly, some of my observations and opinions may seem awkward, or to invite conflict among vested interests, but considering all of the possibilities is necessary to move forward. Einstein said, "We cannot solve our biggest problems with the same thinking we used to create them.

I am also of "somewhat mature years" and have used larger than normal type sizes. This aids legibility, adds pages, and makes the guide look more profound.

Contents

Contents (continued)

"<u>Lean Inventory</u>"?

So, just what do we mean by the term "Lean Inventory"... an empty warehouse? Idle forklift trucks? Purchase of a "Star Trek Synthesizer & Transporter" to replace the manufacturing and distribution system? Hardly!

A notorious oil-well fire fighter named Red Adair is reputed to have told a potential customer, "We can manage this problem better, faster and cheaper than anybody... just pick *any two*."

By achieving "Lean Inventory", we hope to get away from fire fighting, and surpass Red... manage the problem better AND faster AND cheaper... *ALL THREE!*

SOME inventory is usually good and necessary to satisfy customer needs, but to qualify as "Lean", it needs to be wastes-free, flowing (materials and product), and scheduled in accord with actual customer demand (ideally "pulled" from a standardized & reliable production / delivery process by the customers).

Some Consultants portray "Lean" as a sort of mystical Eastern philosophy requiring years of meditation and sacrifice to master. Note the term *YEAR$* ? Actually, the basics of becoming "Lean" are quite logical and straightforward. It is only the details of execution and deployment that can seem mysterious without a guide or "Sensei" (Teacher, who has "been there and done that.")

So, just what are these logical steps, and which details of execution are key, especially with regard to inventory management?

10 Steps to "Lean"

1. Understand customers, demand & trends
2. Map, ID & rank opportunities (VSM)
3. Get clean & organized (5S)
4. Eliminate Wastes
5. Establish "Flow"
6. Incorporate "Pull"
7. Integrate Suppliers
8. Integrate Distribution
9. Makes "wins" standard
    (*NO bureaucracy!*)
10. Continuously improve

Problem-Solving Tools
· Lean Product Design (DFx)
· Concurrent Engineering (3P)
· Mistake-proofing
· RRCA
· 5S
· 5-Why's
· QCPC
· VSM & BPM
· Setup Reduction
· Heijunka
· Rate-Based Planning
· Line Balancing

Integrated Lean
Systems Deployment ®

Lean Step 1)   Understand Demand

A lean system is focused on and matched, in time, to customer demand... Everything happens to customer specification, and "just in time", so being able to define that demand, both qualitatively and quantitatively, is crucial. What product families, models and options will be wanted and when? Not sure? What do history and market trends suggest?

Is forecasting (the "F-Word") an issue? If so, do you measure its accuracy... and if not, how do you expect to improve or at least intelligently apply adjustment ("Kentucky Windage")? Do you have "PBOMS" (Planning Bills Of Material) to help "explode" models and options from (easier to predict) product family demand?

Do you know historic customer-desired shipment or delivery or lead-time requirements?   If so, making your production cycle time shorter (lean manufacturing & distribution) might allow confusion-free "building-to-order" with no finished-product inventory. Plus... even if make-and-deliver cycle time is greater than *SOME* demand, knowing how much "some" is defines the upper limit to needed finished product inventory. NOTE: Historic production or shipping data is practically useless because it contains all historic "lateness" and customers' "patience factor". The goal should be to ship or deliver when customers actually want that.

Lean Step 2)   Map and Identify Opportunities

Value stream mapping depicts and quantifies the performance (wastes and flow) of the entire process... end-to-end... order entry, credit check, planning & scheduling, procurement, production, packaging, distribution & delivery... so that "choke points" (wastes and delays), and their relative improvement priority become clear.

How to do a VSM is beyond the scope of this work, but an excellent first reference/guide is "Learning to See", a Lean Enterprise Institute guidebook by Mike Rother & John Shook.

Perhaps my greatest learning, from leading literally dozens of VSM sessions, is that the greatest time & money savings opportunities are often *beyond* the production floor... in the information-handling, procurement, distribution or product-design arenas!

But even if your operation has NOT (yet) constructed VSM's, all is not lost. Odds are VERY good the key opportunities and priorities are "communal knowledge" if you ask the right people and ask the right questions. The right people are the ones who do the day-to-day work of planning, procurement, and getting product on its way to customers. The right questions include:

- If somebody didn't provide daily work orders, how would you know how to best utilize your knowledge and skills in meeting customer needs?
- What prevents you from doing your job immediately and correctly on the 1st attempt?
- Where in our processes are we wasting time and money that would be better spent to satisfy customers and better compensate employees?
- When Management asks for things to be done "better, faster and cheaper", what are they missing? What, specifically, don't they seem to understand?

Lean Step 3)   Get Clean & Organized (5S)

It is very hard to see wastes and choke points (bottlenecks) if the workplace is totally chaotic and jumbled. Whether in the office, production floor or warehouse, "tidiness is next to Godliness". There should be a place for everything, everything in its place, and nothing unnecessary taking space or creating obstacles to visibility, movement and material flow. The acronym "5S" refers to 5 Japanese terms beginning with "s-sounds", but which loosely translate to:
- Sorting: Identify what's needed, what isn't, and get rid of the latter... ruthlessly!
- Storing: Identify the best place for what's needed, put it there, and *label* the space or location.
- Shining: Clear away dust, debris & litter, Get product presentable for customers, facilities suitable for productivity, and equipment cleaned enough to easily determine readiness or maintenance needs.
- Standardizing: Make the "S's" the "norm"... the accepted way of working..., and make all needed tools or aids readily accessible.
- Sustaining: Establish ownerships, management reviews, metrics and rewards or reminders for sustained performance. Also: Create a FORMAL time allowance for people... before starting, mid-day and before quitting production for the day.

This is much more than simple "housekeeping". Done properly, it can drastically reduce required work area and people movement, speed the flow of information & material, and make next-improvement-opportunities highly visible and obvious.

I recall escorting a Japanese Consultant onto a supplier's production floor, and having his translator say, "I see a combined land-fill and materials warehouse. Where is the factory?". He was right, and simply correcting that situation worked productivity and capacity miracles. (He also asked how many people worked there... to which the Plant Manager replied, "About half.")

Lean Step 4)   Eliminate Wastes

Searching for "MUDA"

The 7-Deadly Process Wastes

Overproduction	**(Building ahead or JIC vrs. JIT)**
Transport	**(Moving things too far or more than once)**
Delay	**(Waiting for materials, machines, next workpiece)**
Processing	**(Doing more than req'd)**
Inventory	**(Any, unless actually being worked on)**
Motions	**(Unless minimum req'd to add value)**
Defects	**(Including "wrong first time" and "try agains")**

Overproduction: Building or procuring more than the customer wants (yet) is clearly wasteful from an inventory standpoint. The excess consumes capacity and materials for production that *IS* wanted, takes space, ties up capital monies, risks obsolescence, and makes the resolution of quality issues more difficult. Regardless of customer order levels, apparent economic batch sizes or historic scheduling/buying practices, DO NOT OVER-PRODUCE!   The one possible exception to the rule is when pre-season production is needed to cover short capacity during peak demand periods...but even then, it could be better either to cap orders or to streamline operations, reduce batch sizes, and boost peak capacity.

Note that procurement timing is as key as production timing. Global sourcing may *SEEM* to have cost advantages, but beware of lot sizes and lead times dictated by long-distance transit needs.   "Just-In-Time" almost always remains the best/fastest/cheapest *total*-cost solution, and favoring suppliers within a 1-hour delivery distance usually pays dividends.

This is difficult to prove without full "Activity-Based Accounting", but the probable lack of that is Finance and Accounting's problem. Don't let them make it yours!

Needless or Redundant Transportation: An obvious waste of time and resources, but with inventory management ramifications that are not always "painless".

Storing materials off-site? Why not reduce days-of-supply from reliable suppliers and use any on-site space to keep "every day" items? This could drastically reduce "shuttle" costs.

Taking multiple weekly deliveries from 10-12 local suppliers? Why not contract a single trucker to make a daily loop route or "milk run", and only need to deal with one delivery per shift?

Why not spend a a few hours intercepting and challenging every truck and container that arrives… asking, "How could we have received this material in a smarter, more timely & more cost effective way?"

Delays: Whether considering material, product or information, "time is money" and delays are deadly. Late deliveries require bigger safety stocks. Batching of orders or production creates delay and "in-process" inventory by definition, incurs extra cost, and slows customer response. In a "lean" operation, material and information never stop moving except to have actual value added, once they enter the process.

Over-processing: This doing more than is actually required. In a manufacturing environment, it might take the form of painting or polishing a surface that cannot be seen after final product assembly. In the Inventory and materials arena, it might include having to remove padding from individually-wrapped components, full-wood crating of products that are being loaded to containers, or staging product from a delivery instead of putting it away immediately (all that's needed is bar-coding and advance-bills-of-lading so the crew can "make a hole" where it will be needed).

I know one manufacturer who pays large sums to have his company logo printed on all 4 sides of his product containers, but these are immediately removed and scrapped by the installing contractor once on the job site, and are *never seen by the end customer*. This inflates both product and inventory holding costs, but adds no real value… MUDA!

Inventory!: In theory, material or product not actually moving or being worked on is non-value-adding and wasteful (the customer would object to the cost if he or she knew). We all know that SOME inventory is inevitable, because different processes have different cycle times, and "buffers" enable them to work as a system. This does *NOT*, however, excuse weeks or months (even years) of supply of materials, components, supplies, or finished products.

I recently (March of 2009) bought "new" _Japanese_ tires for my car (a 2004 model), and discovered DOT date stamps indicating they were made in "0702" (week 7 of 2002). What is wrong with this? How much of the exorbitant price I paid was due to inventory holding costs for 7 years? What's the quality like after so much aging time? MUDA!

Non Value-Adding Motion: Doing anything more than once is wasteful... MUDA! A classic example in the inventory arena is "burying" one product with another (due to space limitations or lack of assignments). Every need for a (buried) unit requires non-value-adding moves of other material, and inflates holding costs.

Similarly, trucks or containers need to be reverse-loaded... first-delivered items loaded last. Random loading creates delays and incurs NVA motion or handling, which in turn risks damage or "shrinkage".

Location strategies and systems will be examined in a later section of this guide, but clearly, *where* material is held can make or break handling efficiencies, and difficulty finding the needed items is an automatic *DELAY* (Muda).

Defects: This clearly includes "poor quality that escapes" to the customer, but also includes the waste of time and/or material anytime an intended action fails on the first attempt ("turnbacks and/or rework"). From an inventory standpoint, it includes any material or product damage, type or quantity mistakes in order processing, AND any information-based mistakes such as with receiving/shipping/invoicing/location records

"Value-Adding"?: In summary, a "waste" or "muda" is anything that does not "add value"... anything the customer would object to paying for if they knew about it. The exception is actions or costs that are "NVA but necessary or mandated", such as verifying shipment weights or inspecting & testing items for code compliance. This begs a "logic-tree" and matching "Lean Strategies":

Value Identification

Eliminate the abnormal and the unnecessary, streamline the non-value-added-but-necessary, ensure flow for the value-added

Lean Step 5):   Establish Flow

Reducing lot sizes and eliminating wastes and delays will dramatically improve the flow of materials and information, but additional measures will still usually be needed.

Physical layout is an obvious element. Organizing people and equipment by product group or mission creates better flow than organizing by technical function or activity. Likewise, positioning people, equipment and storage to minimize moves and transports will aid flow dramatically. Finally, consider co-locating functions with shared missions. Housing Engineering or Order-Entry in different locations from Production hurts flow. Info-Technology does *NOT* compete with "eyeball-to-eyeball" connections.

All processes have natural "bottlenecks" or "choke points" because certain steps take longer than others and cannot be easily subdivided. Reducing batch or lot sizes helps, but where step-subdivision proves awkward (e.g. run-testing or painting), it may be necessary to completely duplicate that step or work station and split production across two or more parallel paths.

Also pay attention to scheduling logic. A "perfect" final-assembly-based scheduling system is doomed if a fabrication or sub-assembly operation is the slowest element in the overall process. A "correct" scheduling system focuses on the slowest component or step, *AND*, continuous improvement effort is focused on that bottleneck.

Remember, also, the "Law of Constraints" (Resolving one bottleneck only results in the next one taking its place).   This is why Value-Stream-Mapping helps so much.

Ultimately, flow is only optimized when we implement Step 6)...

Lean Step 6):   Incorporate "Pull"

Ideally, Customers draw from (small, temporary) inventory, and Production replaces what is withdrawn. Likewise, Fabrication, Sub-Assembly and Suppliers simply replace components consumed by Final Assembly and/or Distribution.

In this way, inventories are kept small, "forecasts" become less critical, and everyone is simply replenishing consumed items rather than chaotically producing material that may or may not be needed.

In the real world, customers behave erratically and "change" their preferences, so inventory (assortment and level) tends to be larger than theoretically necessary.

The point is to target and enable "pull systems" wherever possible... including at the finished-product warehouse or store-room!

If an item is infrequently wanted, offer it on a *strict* build-and-ship-to-order basis... NO stocking!

Similarly, it is not uncommon that "50%" of the produced and offered models represent only "5%" of sales. Here, "weeding the garden" and *enforcing* intended product super-cessions is vital (see the prior comment re infrequently-requested items).

Lean Step 7): Integrate Suppliers

Help key suppliers implement 5S, Waste Reduction, Flow and Pull (unless they are already well on the path to "Lean"). If we do not, they will become the ultimate "bottleneck or choke-point". If this seems like "investing in somebody else's business", think objectively about the alternatives, and switch any suppliers who are also key to competition.

As with Quality: If 75% of your challenges are supplier-based, at least 75% of your improvement resources & people should be supplier-focused.

Lean Step 8): Integrate Distribution

ALL parts of the value chain need and will benefit from becoming "Lean", but leave Distribution for last. Production reports to "Headquarters", and Suppliers want to please Production, but Distribution tends to think and act like the "Customer". Its needs must be satisfied and (product-source) problems resolved before internal improvement can be discussed or motivated.

Lean Step 9): Standardize "Wins" and Successes

Anytime a change has positive results, do what's necessary to make it the "new and official way of working". Dismantle old policies, measures, instructions, practices and incentives, and replace them with the new.

Reward compliance as warranted and deserved.

This is sometimes "painful" if we have ISO, Baldridge or similar certifications, which are built on elaborate process documentation that will need revision, but "Just Do It!"

At the same time, work hard at avoiding new bureaucracy and extensive documentation. Both are frustrating and intimidating to future change and improvement.

The attributes of a beautiful 20-year-old are very different, but *not* more wondrous, than the charms of a 50+ beauty. Allow for positive change and make it a goal rather than a burden.

Lean Step 10): Continuously Improve

Getting "Lean" is an ongoing journey, never really concluded. When one problem or constraint is removed, another inevitably takes its place. The magnitude of the challenge may wane, but, "like being on the railroad tracks... if you stand still, you will get run over."

New customers, products, services, processes, suppliers, codes, competitors and economics are an ongoing driver of improvement needs, which is why we hear the seemingly illogical advice: "If it is not broken, break it!"

Production Inventory Management Fundamentals

Some inventory is necessary to manufacturing and to meeting customer lead time expectations, but it is also a major form of working capital, and can significantly detract from cash flow, and profitability. Inventory can also disguise supply-chain process problems such as on-time delivery, long lead times, quality, etc.

Overall, there are three types of production-level inventory...

1. Inbound or on-site raw materials and components
   - In-transit
   - Needed and ready for imminent use
   - Surplus (more than immediately needed)
   - Damaged or scrap material
   - Obsolete material & components
2. Actual work in process (no longer "available", but unshipped)
3. Outbound materials (finished goods, accessories and service parts)

Clearly, a well-developed "P-I" (Production Inventory) management strategy includes specific policies and practices to maximize "flow"... to balance work and keep both materials & products moving.

The purpose of what follows is to outline specific actions and procedures to best-manage inbound and ready-to-use materials and components. The core elements of these actions and procedures are well-known and/or the subject of separate guides, so emphasis here will be on combining and linking these into **reliable systems and processes**... an inventory management framework that can be taught and relied upon to achieve needed, key objectives:

- Satisfy customer requirements, especially lead time and promised shipping date
- Avoid material shortages (keep lines operating, without schedule and work-sequence juggling)
- Minimize surplus, unnecessary or premature material acquisitions
- Account for material status and availability at all times
- Provide Management with meaningful and timely measures of both process performance and operational impact.

Wide differences in customer requirements, product type & complexity, supplier base and scope or size of manufacturing units mean there is no single, best solution for all situations. The challenge for each Procurement & Materials Management team is to understand the essence of what follows, and to appropriately determine how the intents and purposes can be best achieved for their site or unit.

Exhibit 1: A "Strongly-Recommended" Process & Plan

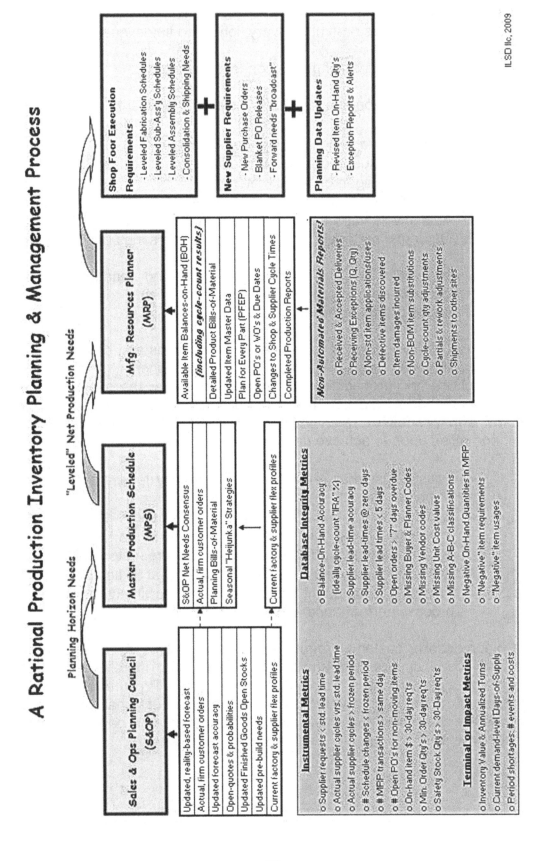

20

A comprehensive overall "Production Inventory Management Process" is shown in Exhibit-1. , and is quite straightforward. The key steps include:

1. A "Sales & Operations Planning" meeting is regularly conducted to review customer demand, market trends, and to reach consensus on the specific product production for the upcoming planning and scheduling period.
2. This proposed production then passes to the "Master Production Scheduling" process for a reality-check and alignment with capabilities (plant and supply chain).
3. The "Master Schedule" is then passed through the "Manufacturing Resource Planning" System (MRP computer), which determines material & component needs, item or task lead times, needed event sequences, current materials availability, and planned arrivals, then generates the next (3) steps or outputs...
4. Shop Floor Scheduling and Execution Requirements
5. Supplier Scheduling and Execution Requirements (Procurement), and...
6. Forward-Planning Data & Info Updates (including exception reports)

The **"secret to success"** depends greatly upon well-developed data & info inputs to each process step, upon discipline within the process (rules-based decisions and no shortcuts or "end-runs"), upon execution-to-plan (no ad-hoc or informal deviations), and upon carefully designed measures of both process functioning and operational impact.

Accordingly, this guide emphasizes key process inputs (with examples where appropriate), suggested metrics (and "non-metrics"), and some policies/tools/strategies with recognized strong impact on production inventory levels.

The Sales and Operations Planning Meeting (SOP)

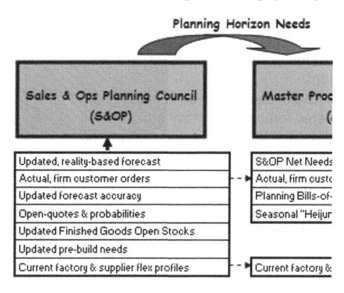

Exhibit 2A

Companies with apparently successful material planning control systems generally schedule and hold a regular "SOP" meeting, conducted in advance of their periodic MPS/MRP cycle (see Exhibit 2a).

The principal objectives of the SOP include a *reality-based update and consensus* re planned or forecast demand… during the immediate planning period and beyond. This enables the sales and order management team to make realistic promises, enables the materials procurement team to acquire and maintain adequate but efficient component stocks, helps operations to plan work schedules and staffing, provides better accuracy and credibility to supply-chain requirements broadcasts (best-estimates), and recalibrates management expectations for performance versus "plan".

This idea of "reality-based" planning & scheduling avoids the risk of simply building to a projected business plan or forecast that may have been assembled months prior.

Participation should be as cross-functional as possible. Key members include Sales, Order-Entry, Production Planning, Scheduling, Purchasing, Finance and Operations Management.

Meeting frequency is, at minimum, linked to the schedule for MPS updates, and is best conducted using an in-person format (versus teleconference), so that discussion is not as time-constrained and so that "body-language" can contribute to communication.

The quality and value of SOP discussions is closely linked to the type and quality of information available to the team. Some of these inputs are logical, but others require training and development:

- Firm customer orders
- Major open quotations, closure probabilities, and expected timing
- Extended forecast, updated for market trends and competitive actions
- Updated seasonal pre-build plans (if any)
- Updated forecast accuracy analysis
- Factory and key-supplier response profiles

These last 2 are the least traditional.

Forecast accuracy analysis simply consists of tracking and reviewing how well 30, 60 and 90-day demand forecasts have compared to the eventual reality. By providing this data and feedback to sales and others involved in making the forecast, objective process-review can take place, and accuracy can improve. In addition, Operations can objectively estimate possible forecast error.

__Response profiles__ provide quantified insights to capacity and flexibility limits over time, for both the factory and for key suppliers. For example, a key supplier's response profile may tell the planning team that performance boundaries usually include:

- Up to 20% supply increase over normal with minimal warning, *but* subject to raw material availability (e.g. certain types or sizes of sheet metal)
- Up to 40% increase over normal with 10-days advance notice (weekend hrs needed)
- Up to 100% increase over normal with 30-days advance notice, *AND* a commitment to at least a 60-day demand-level continuation (second shift needed).

By knowing these limits, the SOP team can avoid making unrealistic plans and promises, know when and who to consult if greater response flex is needed, and make decisions far enough in advance for key contributors to support needs.

The final product of the SOP is a **"Net Planning-Horizon Needs"** list, adequate to compose the next-cycle's Master Production Schedule AND to objectively assess long-lead-time procurement needs. Here, three concepts are key:

1. Requirements updates should assume and honor a reasonable "frozen schedule period" in the immediate future. Significant changes to a near-future production schedule most often lead to chaos because of lead time and response profile violations, the high probability of materials shortages, and the "domino effect" on all operational processes. Even Toyota observes and recommends at least a 2-week "frozen schedule".
2. Requirements should also anticipate and support the practice of "scheduling- heijunka" or load-leveling. To the greatest extent possible, scheduling should be done in a way which spreads and levels demand over the planning period, and which incorporates increases and/or decreases as gradually as possible. This avoidance of batching and large or rapid quantity swings promotes quality, productivity, and on-time-delivery. It also minimizes the need for expediting and overtime.
3. Long supplier lead times, long transit times and load leveling may require temporary loading of demand *forecasts* (imaginary or virtual orders) to the schedule (MPS) *beyond* the firm-schedule period. The alternative is usually shortages and surpluses from Buyer/Planner guesswork.

Here is a sample SOP meeting agenda:

Sales & Operations Planning Meeting (SOP)

Participation

Mktg, Sales, Order Entry, Engineering, Scheduling, Procurement, Production, Logistics

Typical Agenda

Review of known, detailed requirements

Customer order backlog
New, firm customer orders

Review of likely, additional requirements & factors

Open quotes (incl. liklihood of closure & timing)
Updated marketing/sales forecasts
Updated seasonal pre-build needs
Market trends and forces
Competitive actions & implications

Analysis of Needs versus Constraints

Group requirements by product or family
Compare to manufacturing plans and limitations
 Work plan: hours, days, weekends, shifts
 Equipment capacity
 3PL implications if any
Compare to supply chain limitations
 Key supplier capacity & lead times (esp. global)
 Key supplier work plans and change times
"Frozen Schedule" maintenance check
New requirements fit to "flex schedule" boundaries

Requirements Concensus

Frozen Schedule Period requirements verification
Defined "next period" requirements (level of detail per MPS team needs)
Updated, rollong forward forecasts (per available detail)

Follow-up assignments

Mktg, Sales, Order Entry, Engineering, Scheduling, Procurement, etc?
Minutes and info distribution

The "MPS" or Master Production Schedule

The Master Production Scheduling step converts the consensus requirements from the SOP into a specific list and sequence of end-product needs and completion dates that the MRP computer program can understand and use to determine materials and resource requirements. (see Exhibit 3a)

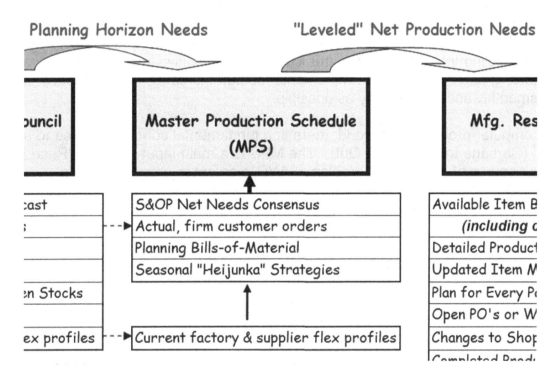

Exhibit 3A

Since the MRP is simply a sophisticated accounting device, the Master Scheduler(s) must supply intelligence and judgment not part of the MRP logic, making this a key role for professionals with solid knowledge and experience of the operation. Among the more challenging considerations are:

- Converting (high-level) requirements to detailed (build-able) unit or sub-assembly needs using historic mix models and "Planning Bills of Material" (see below).
- Accounting for current finished goods inventory (sold or allocated versus available), avoiding over-stocks and supporting first-in-first-out "FIFO" stock management.
- Loading enough forecast-based requirements beyond the immediate, frozen-period" or planning window to enable the procurement of long lead-time items, and to enable meaningful "broadcast" data for suppliers. This requires very close thought where the product is complex (numerous options or varieties) and the forecast is "vague" (see discussion on PBOMs).

- Preserving the integrity of previously "frozen" schedules, or at least quizzing those impacted about the acceptability of any changes (including critical suppliers). Here it is helpful to have some basic flexibility rules or guidelines agreed in advance… an acceptable amount of quantity variation, type of unit, specification variance, sequence and/or batch size adjustments, etc.
- Understanding and honoring the factory and supplier response profiles, and avoiding predictable disappointments.
- Understanding and supporting the practice of "**heijunka**" or load leveling… scheduling in a way that creates consistent, repetitive requirements for all task groups, and which implements output increases or decreases as smoothly and as gradually as possible.

In the computer-programming world, there is a fundamental concept referred to as "GIGO" (Garbage In = Garbage Out). The MPS is a main input to the MRP, so is at once a source of most output problems, *AND*, a secret to preventing them.

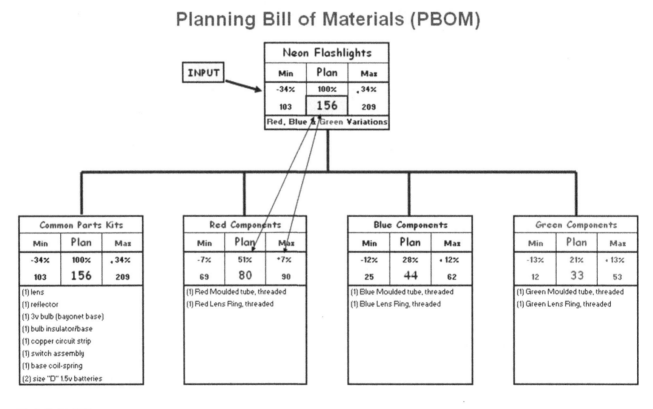

Planning Bill of Materials (PBOM)

Exhibit 3B

Planning-Bills-Of-Material (PBOMs) allow the application of statistically derived product size or option probabilities to overall demand estimates. By forecasting at high or family level (e.g. number of all units versus number by size and detailed type), forecast accuracy can be dramatically improved. Done properly, a PBOM will also predict a minimum and maximum range of requirements at material-planning levels of detail… making it much easier to have what is ultimately needed without buying and holding unnecessary (improbable) quantities "just in case". PBOM development does require a reasonable amount of historic data, and does assume fairly consistent customer preferences. The Details of how these models are composed is beyond the scope of this guide, but a simple model is offered in Exhibit 3b, and an excerpt from a real-world example is depicted in Exhibit 3c. Detailed PBOM training is available from ILSD llc, and is described in my forthcoming "FAST" (lean Forecasting And Scheduling Tips) guide.

8

Monthly Family Demand Projection - [200]

Item	% Mix	% Var	Material Kit Requirements Min	Plan	Max	3 Sigma Risk-0% Max	2 Sigma Risk-3% Max
All Solar AC Units (Family)	100%	86%	28	200	372	315	257
All Coils (Major Sub-assembly)	100%	86%	28	200	372	315	257
Size 1	9%	170%	0	18	48	38	28
Size 2	3%	573%	0	6	39	28	17
Size 3	9%	165%	0	19	49	39	29
Size 4	9%	204%	0	19	57	45	31
Size 5	3%	307%	0	7	28	21	14
Size 6	6%	141%	0	13	30	24	18
Size 7	4%	465%	0	8	47	34	21
Size 8	6%	340%	0	11	51	38	24
Size 9	2%	314%	0	4	16	12	8
Size 10	10%	215%	0	21	66	51	36
Size 11	8%	219%	0	16	52	40	28
Size 12	11%	204%	0	21	64	50	35
Size 13	3%	494%	0	5	32	23	14
Size 14	8%	353%	0	16	71	53	34
Size 15	3%	627%	0	5	39	28	17
Size 16	5%	405%	0	11	55	41	26
Cabinets (Major Sub-assembly)	100%	86%	28	200	372	315	257
Frame 2 (sizes 1&2)			0	24	87	66	44
Frame 4 (sizes 3&4)			0	37	107	84	60
Frame 6 (sizes 5&6)			0	19	58	45	32
Frame 8 (sizes 7&8)			0	20	97	72	45
Common Parts, Frames 2-8			0	100	349	267	182
Frame 10 (sizes 9&10)			0	25	82	63	44
Frame 12 (sizes 11&12)			0	37	116	90	64
Frame 14 (sizes 13&14)			0	21	103	76	48
Frame 16 (sizes 15&16)			0	16	95	69	42
Common Parts, Frames 10-16			0	100	396	298	198
<Cabinet "H" or "V" mounting options estimate - all sizes			0	200	745	565	380
Fan Decks (Major Sub-assembly)	100%	86%	28	200	372	315	257
Frame 2 (sizes 1&2) (5 fan rpm options)							
Frame 4 (sizes 3&4) (5 fan rpm options)							
Frame 6 (sizes 5&6) (5 fan rpm options)							
Frame 8 (sizes 7&8) (5 fan rpm options)							
Frame 10 (sizes 9&10) (9 fan size/rpm options)							
Frame 12 (sizes 11&12) (9 fan size/rpm options)							
Frame 14 (sizes 13&14) (9 fan size/rpm options)							
Frame 16 (sizes 15&16) (9 fan size/rpm options)							
Focus Item 914 mm Fan Blade Assemblies (avg 5	99%	125%	0	1151	2589	2114	1625
1240mm Fan Blade Assemblies (avg 4	1%	841%	0	8	71	50	28

(Fan Decks note: "same as cabinets")

Exhibit 3C

MRP: Manufacturing Resource Planner (usually automated)

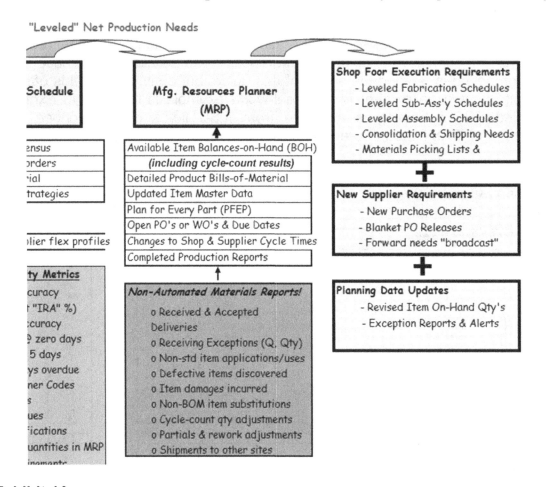

"Leveled" Net Production Needs

Schedule

ensus
orders
ial
trategies

lier flex profiles

ty Metrics
curacy
"IRA" %)
ccuracy
zero days
5 days
ys overdue
ner Codes
s
ues
ications
uantities in MRP

Mfg. Resources Planner (MRP)

Available Item Balances-on-Hand (BOH)
(including cycle-count results)
Detailed Product Bills-of-Material
Updated Item Master Data
Plan for Every Part (PFEP)
Open PO's or WO's & Due Dates
Changes to Shop & Supplier Cycle Times
Completed Production Reports

Non-Automated Materials Reports!
o Received & Accepted Deliveries
o Receiving Exceptions (Q, Qty)
o Non-std item applications/uses
o Defective items discovered
o Item damages incurred
o Non-BOM item substitutions
o Cycle-count qty adjustments
o Partials & rework adjustments
o Shipments to other sites

Shop Foor Execution Requirements
- Leveled Fabrication Schedules
- Leveled Sub-Ass'y Schedules
- Leveled Assembly Schedules
- Consolidation & Shipping Needs
- Materials Picking Lists &

New Supplier Requirements
- New Purchase Orders
- Blanket PO Releases
- Forward needs "broadcast"

Planning Data Updates
- Revised Item On-Hand Qty's
- Exception Reports & Alerts

Exhibit 4A

This (usually automated) step in the process does all the calculating. The master production schedule is translated or "exploded" by the product bills-of-material to determine gross material and components requirements over time. Balance on hand is updated for completed production (per the BOMs) and/or supply-item requisitions, and new fabrication or purchasing needs are determined after allowing for already- scheduled work orders and deliveries.

The question of MRP full-update frequency is a common one. The procedure does take time (sometimes several hours) and requires supervision by skilled IT professionals, so running a full update really only adds value if and when the MPS has changed or has been updated. At the same time, daily or nightly processing of all materials transactions and completed-production reports (relatively quick) is important to having accurate BOH data (net useable item quantities) available.

The **(3) main results** of this step are:

1. Shop-floor scheduling and execution (work) schedules. These typically include materials "pick lists" and "routings", and are (ideally) sequenced in time to "level-load" all the working groups.
2. Supplier scheduling & execution (needed delivery) schedules. POs or Releases can be automated or manual, but should also reflect "load-leveling" measures.
3. Forward-planning data & information updates… including BOH maintenance, planning exception reports, etc.

Clearly, the accuracy and *timely availability* of information inputs are critical, and high-performance management processes need to include methods to monitor both dimensions.

System Bill-Of-Materials (BOM)

BOM accuracy is vital! Unneeded items or over-stated quantities lead to unnecessary purchases or fabrication. Missing items or under-stated quantities lead to shortages. Care is needed to reliably track the needs, usage and availability of hard-to-quantify (or un-quantified) materials such as paint, sealants, masking tape, etc.

Item Master and "PFEP"

A part of the database called the "Item Master" contains all the key item parameters needed for either purchasing or fabrication shop-ordering (source, lead time etc), BUT typically lacks information needed to manage physical logistics planning and work, so is ideally based on a (usually separate) "PFEP" or "Plan-For-Every-Part". A simple example of this is shown in Exhibit 4b. A well-managed system tracks BOM error reports and resolutions, and this function is a clear part of the Sustaining Engineering group's performance measures.

Plan For Every Part (PFEP)

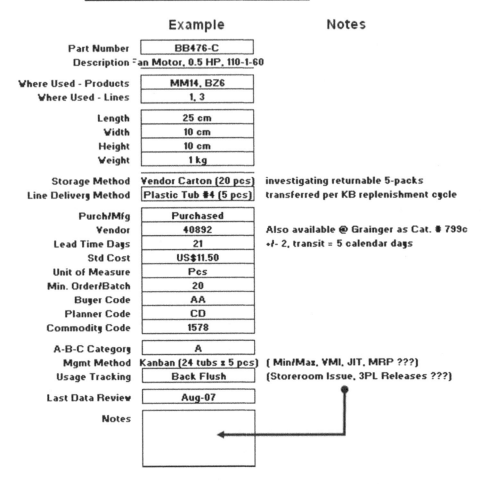

	Example	Notes
Part Number	BB476-C	
Description	Fan Motor, 0.5 HP, 110-1-60	
Where Used - Products	MM14, BZ6	
Where Used - Lines	1, 3	
Length	25 cm	
Width	10 cm	
Height	10 cm	
Weight	1 kg	
Storage Method	Vendor Carton (20 pcs)	investigating returnable 5-packs
Line Delivery Method	Plastic Tub #4 (5 pcs)	transferred per KB replenishment cycle
Purch/Mfg	Purchased	
Vendor	40892	Also available @ Grainger as Cat. # 799c
Lead Time Days	21	+/- 2, transit = 5 calendar days
Std Cost	US$11.50	
Unit of Measure	Pcs	
Min. Order/Batch	20	
Buyer Code	AA	
Planner Code	CD	
Commodity Code	1578	
A-B-C Category	A	
Mgmt Method	Kanban (24 tubs x 5 pcs)	(Min/Max, VMI, JIT, MRP ???)
Usage Tracking	Back Flush	(Storeroom Issue, 3PL Releases ???)
Last Data Review	Aug-07	
Notes		

Exhibit 4B

Operations and Supply-Chain Lead Times

Supplier or fabrication lead times are a COMMON source of calculation error. Some can be seasonal, and all should include TOTAL acquisition cycle times... work or purchase order prep & release, shop or supplier response time, transit, receiving, repackaging etc.

Verifying lead times is best accomplished by writing an MRP query-report that compiles actual order dates versus receipt or completion dates for 2-3 months, and compares this to nominal lead times in the Item Master.

Accounting for Used or Allocated Materials

Accurate & timely completed-production reports are an essential ingredient for "back-flushing"... using the product BOMs to calculate and debit materials usage from available balance-on-hand. This SHOULD be straightforward, but requires reliable methods to closely monitor and account for:

- "Almost completed" production ("partials"): If back-flushed, someone needs to credit the BOH of the (unused) missing items until the products are completed, then again debit those balances. If not back-flushed, delay-time needs to be closely watched, since true component availability will be less than assumed by the MRP.
- Product that is on "quality hold" or in "rework" are similar in effect, and in the kinds of management required.
- Final Test duration can also create BOH concerns. When long (e.g. overnight), it may be preferable to back-flush the production first.

Manual or "Floor" MRP Transactions

Finally, but by no means least important, are the accurate and *timely MANUAL-entries to the MRP* system of materials transactions that an automated "backflush" cannot and will not account for:

- Accepted deliveries (Receiving)
- Receiving exceptions (quantity, quality deviations)
- Supply item issues or requisitions (non-BOM item uses like paint & tape)
- Defective items
- Damaged items
- Material substitutions and/or "KBI" design revisions (Known But Incomplete)
- BOH adjustments during physical inventory events
- Partials and rework adjustments (described above)
- FSOs ("Field-Shipping-Orders") or transfers to other sites
- "Cycle-Count" BOH adjustments.

All will impact balance-on-hand accuracy, plus defects & damage tracking are key to quality-improvement.

This requires commitment, discipline and accountability on the part of floor Supervisors and assigned staff. Well-managed systems track and report the frequency of manual transactions by type and/or responsible group. There may not be a right or wrong, good or poor number, but changes to the level or upward/downward trends can suggest "opportunities" and compliance levels.

Full physical inventory counting may be needed per the requirements of financial reporting, while "**Cycle-Counts**" are individual-item BOH checks and adjustments based on item type and value. Advanced MRP systems can manage this process... specifying usage and/or value-biased cycle-count lists... that if followed, can provide 98%+ BOH accuracy and even eliminate the need for full physical inventory. At very least, item cycle-counting should be done whenever:

- A shortage occurs
- Material is not found in the assumed or assigned location

- The MRP BOH seems inexplicably high (usage tracking problem)
- The physical, floor quantity seems suspiciously large or small
- Purchase Orders are late for any reason (requested delivery is under agreed lead time)
- Major Receiving Exceptions are posted (quantity or quality shortfalls)
- Supply-chain interruptions risk late deliveries (port or transport disputes etc)

MRP "outputs" are listed above and largely self-explanatory, but well-managed processes are typically strong in three specific areas: schedules & sequence "leveling", supplier "broadcast" generation, and "exception reporting".

Schedule and work sequence leveling may not be within the (automated) capability of some MPS/MRP systems, so a manual or Excel®-assisted step may be needed between the standard MRP work/procurement plan and actual releases to floor or suppliers. This additional step is generally worth the time and effort because of its positive impact on quality, productivity and on-time-delivery. In addition, however, it supports consistent materials flow and reduced WIP inventory levels. This is particularly true when production lot sizes and/or supplier minimum order quantities have been minimized. (see Exhibit 4c)

Leveling the load for Production and Suppliers

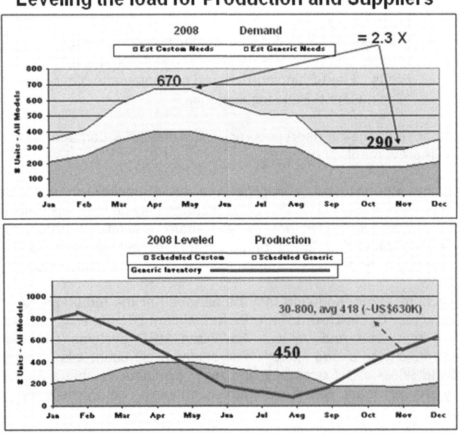

Supplier effective-capacity, lead time and on-time delivery are all typically improved when MRP data is compiled or enhanced to create meaningful "**Supplier Broadcasts**"... best estimates of forward requirements including *and beyond* the firm-order period... especially if linked to agreed supplier response profile data (described earlier). Most MRP systems require some form of "data post-processing" (additional software) to generate these broadcasts. Excel® can do the job in simple situations. One example of a supplier broadcast is depicted in Exhibit 4d. A caution: material usage, BOH and broadcast data should be maintained and communicated, even for items managed using VMI (vendor-managed-inventory), kanban, FIFO-lane, min/max or consignment. The data is key to planning, system-sizing, and invoice-auditing.

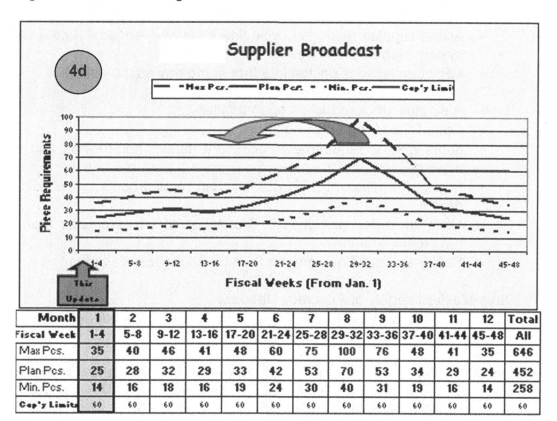

Month	1	2	3	4	5	6	7	8	9	10	11	12	Total
Fiscal Week	1-4	5-8	9-12	13-16	17-20	21-24	25-28	29-32	33-36	37-40	41-44	45-48	All
Max Pcs.	35	40	46	41	48	60	75	100	76	48	41	35	646
Plan Pcs.	25	28	32	29	33	42	53	70	53	34	29	24	452
Min. Pcs.	14	16	18	16	19	24	30	40	31	19	16	14	258
Cap'y Limit	60	60	60	60	60	60	60	60	60	60	60	60	60

Fiscal Week	1	2	3	4
Max Pcs.	8	8	9	10
Plan Pcs.	6	6	6	7
Min. Pcs.	3	3	4	4

Standard Lead Time:	14 Days
Latest L/T Performance	12-16 Days
Planning Lead Time:	16 Days
Flow Strategy:	Kanban
Flex Strategy:	Reserve plan cap'y
Packaging:	5 pcs/container
Capacity Limits:	60 pcs/mo.
Latest LCD Rate:	100,000 ppm

Date: 28 Dec. 1999
Material: Blue Tubes and Tops
Supplier: Sloe Manufacturing

Remarks: Recent late deliveries_ Please advise corrective action plan within 5 days.
Quality unacceptable_ 100,000 ppm = 10% defects!
Demand may exceed your capacity in wks 21-36_ Prebuild may be required.

MRP _**exception reports**_ are a powerful management (and inventory shortage/surplus avoidance) tool. Some are pre-programmed, but require activation. Others require the creation of "queries". All require time and care to ensure that the MRP has all the data needed to perform the analysis and create the reports. Some of the most useful reports include:

Shortage-Avoidance Alerts…
- Potential shortages (inadequate or marginal BOH)
- Overdue, scheduled deliveries (still un-received after due date)
- Items with "zero" lead time settings
- POs with requested deliveries inside of agreed supplier lead times
- Schedule changes inside of the alleged "frozen period"
- Actual supplier response cycle times (delivery-request dates) versus system lead time settings
- MRP transactions posted late (not same day as the event)

Low-risk surplus stock reduction opportunities…
- Open POs for non-moving items
- Items with no requirements or recent usage (non-moving and slow-moving)
- Items with apparent BOH greater than requirements in the next "30" days
- Minimum order quantities or fabrication lot sizes greater than "30-day" requirements
- Safety-stock settings greater than lead-time requirements

Item-Master Integrity or Accuracy Needs…
- Items with missing planning parameters (buyer code, planner code, vendor code, lead time, unit cost, etc.

The studies and reports can be pre-programmed and appropriately scheduled to minimize both user effort and system calculation time, but, _all Buyers, Planners and Inventory/Materials Managers should have training in how to write MRP queries, and access to that capability within the system._ Separate software (e.g. "IQR®") is also available to provide or supplement the MRP query and analysis toolkit.

Measurement Types and Frequency

Instrumental Metrics	Database Integrity Metrics
o Supplier requests < std. lead time	o Balance-On-Hand Accuracy
o Actual supplier cycles vrs. std. lead time	(ideally cycle-count "IRA" %)
o Actual supplier cycles > frozen period	o Supplier lead-time accuracy
o # Schedule changes < frozen period	o Supplier lead-times @ zero days
o # MRP transactions > same day	o Supplier lead times < 5 days
o # Open PO's for non-moving items	o Open orders > "7" days overdue
o On-hand item $ > 30-day req'ts	o Missing Buyer & Planner Codes
o Min. Order Qty's > 30-day req'ts	o Missing Vendor codes
o Safety Stock Qty's > 30-Day req'ts	o Missing Unit Cost values
	o Missing A-B-C classifications
Terminal or Impact Metrics	o Negative On-Hand Quantities in MRP
o Inventory Value & Annualized Turns	o "Negative" item requirements
o Current demand-level Days-of-Supply	o "Negative" item usages
o Period shortages: # events and costs	

As with any core business process, there are three needed types of measurement:

1. Impact or terminal metrics… measures of whether the process is meeting the needs of the business and its customers.
2. Instrumental metrics… measures of whether the process steps are meeting intermediate goals and performing efficiently.
3. Data integrity metrics… indicators of data and input completeness and accuracy.

In the case of the raw and WIP inventory planning & management process, one vital measure of impact or terminal effectiveness is the "Number of item shortages and the resulting hours or units of lost production."

Another is the "Value of owned raw/WIP inventory (owned and on-hand, in-transit, or at a 3PL or supplier), in comparison to the maximum value allowable to achieve the WIP turns goal of the business plan." Here, it is important to acknowledge that Finance will review end-of-quarter levels, but end-of-(every) month averages are more representative of real performance. It is also important to acknowledge that sales/production/inventory can exhibit seasonal swings, so realistic reference targets may need to have a seasonal character also.

Many of the exception reports (above) double as **instrumental metrics.**

- Overdue, scheduled deliveries (often the result of delayed Receiving transactions)
- Items with "zero" lead time settings
- POs with requested deliveries inside agreed supplier lead times

- Schedule changes inside the supposed "frozen period"
- Actual supplier response cycle times (delivery-request dates) versus system lead time settings
- MRP transactions posted late (not same day as the event)
- Items with missing planning parameters (buyer code, planner code, vendor code, lead time, unit cost, etc.

One useful addition is the level and trend of the 30-60-90 day forecast accuracy. Another is the "% of suppliers receiving regular requirements broadcasts".

Inventory "containment" depends mostly on effective planning and procurement management. There are, however, some general policies and operational tools that deserve close consideration because they can also impact inventory levels and "flow" (for better or worse).

Overall Business Unit Metrics can sometimes lead to policies and practices which can adversely challenge inventory control or containment strategies:

- **"Burden Absorption"** is a widely-promoted (Financial) concept and metric based on the idea that the impact of plant fixed costs and indirect labor on product manufactured-cost is minimized by maximizing product production. This is true, *IF* the production can be sold. *Otherwise, there is over-production, which is a waste, and waste NEVER helps the "bottom-line".* The better challenge is "Burden Reduction".
- **"Labor Productivity"** is a useful metric, but if calculated as standard-hrs-earned / actual-hrs-paid, it again becomes "tempting" to maximize production, whether needed or not. A more meaningful measure might be "schedule linearity" (units-produced / units-scheduled), assuming schedules reflect true market needs.
- **"Equipment Utilization"** is like burden-absorption, and risks over-production. What is important is that equipment is available whenever legitimately needed, so more constructive measures might include "unplanned equipment downtime".

It is often the case that apparently poor inventory process performances are unwittingly due to measuring and rewarding decisions or practices that detract from what is really wanted or needed.

Other Key Tools and Definitions

A-B-C Inventory Classifications are intended to focus Buyer/Planner efforts on the most critical items, and manage less critical materials with semi-self-managing methods such as kanban or min/max. Most "traditional" A-B-C systems look primarily at relative item usage rather that at relative usage *value* (which is more to the point where turnover is at issue). If the MRP system offers A-B-C assessment using a relative *usage value* method, it is highly recommended. If the option is not offered, consideration should be given to doing separate analysis. These A-B-C designations should also be reviewed and updated at least yearly, and components for new products should be treated as "A's" until actual usage patterns are established.

Scrap, obsolete items and surplus stock all detract from working-capital performance and cash flow, but in a globally dispersed organization, there can be differing understandings of these terms, related corporate policy and recommended actions. Exhibits 5a, 5b and 5c are offered to help create a common understanding.

Scrap

DEFINITION:
- Damaged and unsaleable items
- Includes dirty items not worth cleaning
- Includes customer warrantee returns
- Includes freight-damaged receipts or returns

POLICY:
- Document and sell for salvage value at the earliest possible moment
- No quarantine or formal review board required
- Remove all manufacturer's nameplates
- Press freight insurers for timely release
- Photograph major items in case of audit or other questions

5a

Obsolete Parts, Materials and Products

DEFINITION:
- No sale or use of the item in the last 12months (24 for svc parts), and...
- No forecasted future demand or sustitution potential, including...
- Mfg. components not needed or offered as a service part
- Includes items superceded by newer items and non-reworkable

RECOMMENDED POLICY:
- Monthly (at least quarterly) reviews & disposals by a cross-functional review board
  (Sales, Engineering, Operations, Service Parts, Finance, Unit Executive)
- Once declared truly obsolete, items should be physically disposed of as soon as possible
- Remove all Carrier or Carrier sub-brand nameplates and packaging prior to disposal

DISPOSAL:
- Clearance sale to the last known purchaser(s) of the item
- Resale to source or supplier (rework possibilities or other customers/users)?
- Sale to non-competing (other-industry) users or outlets?
- Manufacturers' clearance broker?
- Scrap or materials-recycling specialist?

5b

Excess Materials, Components and Finished Goods

DEFINITION:
• All Items not sold or used in the last 12 months (24 for svc parts) and without future need, or...
• Items in good condition and with forecasted sales or use, BUT...
  quantities on hand exceed "necessary or justifiable" levels (see below)

POLICY:
• Quantity on hand should be reduced to "necessary or justifiable" levels as soon as possible
• Ongoing management strategies should be implemented to avoid future surplus accumulation

NECESSARY or JUSTIFIABLE QUANTITIES:
• The maximum historic or forecasted need over the worst-case time to get or manufacture more
  (plus a modest safety stock)
• None, in the case of unused, unneeded and/or obsolete/superceded items having no
  service parts use or requirements

REDUCTION STRATEGIES:
• Sell-back or restocking to the supplier
• Sale to other Can   ories or distributors
• Sale to non-competing (other industry) users or outlets
• Reduced replenishment until "right level" is reached (DON"T "kill" your supplier!)
• Do NOT sell at discount to customers (deflates price and destroys near-future demand)!

5c

Similarly, the causes of **material shortages, surpluses and negative BOH** need to be better understood if inventory problem-solving is required.

Some Common Causes of Material Shortages

Mat'ls used before planned
  Defects
  Emergency Orders
  Unauthorized batching
  Schedule advances
  OEM MRP req't too low
  Svc part taken from mfg. Supply

Late Delivery
  Supplier late
  Fab or sub-ass'y late
  Transporter delay
  Incorrect ship-by date on PO
  On-hold by Receiving Inspection

Incorrect MRP/Planning Data
  Wrong Lead Time
  Wrong On-hand qty.
  Delayed receiving data
  Short shipment
  Bill-of-Material error
  Engr. Change not in system

Late PO or Release
  Poor visual status
  No / late info to whse or buyer
  Purchasing backlog
  Buyer out / no backup
  Oversight / rare transaction
  Late/Missing Backflush

Material Location Problems
  Can't find / misplaced
  Mixed parts delay
  No assigned location

Mislabeled/Unlabeled
  Supplier / Fabr. Error
  Receiving error
  Mat'l Handling Error
  Lost/missing label
  Mixed parts in location
  Label unreadable

5d

38

"X Factors" (Surplus Inventory Causes)

SUPPLIER-BASED
- No or low inventories
- Long lead times
- Min. qty > Need / Lead Time
- Unreliable quality
- Unreliable delivery

MANAGEMENT-BASED
- Batching to "absorb" setups
- Quality buffers
- Seasonal prebuilds
- Shutdown prebuilds
- Relocation Prebuilds
- Equipment downtime buffers
- Forecast uncertainty buffers
- "EOQ" or 1st-cost purchasing
- "Frozen" period > mfg. cycle time
- Policy for labor & burden-absorption vrs reduction
- Low confidence in "MRP" and BOH among buvers and Planners

CUSTOMER-BASED
- Erratic requirements
- Delayed acceptance
- Consignment demands
- Bad credit
- Expect parts beyond warranty

Some Causes of Negative Balance-On-Hand

Miscount during physical inventory (value too low):

Miscount when receiving new deliveries (value too low)

Delay at entering deliveries to the MRP in Receiving (used before "received")

Incorrect item quantity in the MRP Bill-Of-Materials (value too big)

Material substitutions or IPOSA-based uses with delayed or missing system entries

Scrap & defect reporting errors
> Scrapped items are over-counted, calculated balance goes "negative", or...
> Scrapped item is mis-labeled or mis-identified, so the good item's balance goes "**negative**"

Un-transacted Rework
> A batch of Item B's (above) are shortened to become Item A's, but...
> The MRP isn't advised on time (or at all) to reduce the balance of B and increase A's, so...
> A quickly shows a **negative** balance, and B experiences a *STOCKOUT!!*

Duplicate Backflushing
> An item is switched from BOM/backflush to "Min/Max" control or "Supplies" status.
> The storeroom then starts reporting use as the item is issued, but...
> The item's MRP "backflush" switch doesn't get reset to "No", so...
> Calculated quantity gets reduced twice for each item actually issue or used and...
> Calculated balance quickly goes **negative**.

5f

Kanbans and Min/Max systems are increasingly popular, but can become a root cause of both shortages and surpluses if not fully understood and properly managed. Briefly...

- These systems are NOT a means of eliminating inventory. They only serve to *limit* the amount of "buffer stock" between material sources and points-of-use.
- They only work well for items with regular usage. If applied to infrequently used items, inventory will greatly exceed needed and reasonable levels.
- The size of either stock is based on demand over the replenishment lead time. Since both of these factors vary, regular re-sizing is needed to optimize flow and/or avoid shortages.

Outsourcing of component fabrication or sub-assembly operations can streamline operations and reduce the cost of goods sold, BUT, long supplier lead times, long transit times, uncertain quality or unreliable delivery timing can all lead to unwanted safety and cycle-stock inventory increases. Careful source selection and management is critical. "Core Technologies" are a major, separate subject, but in summary, if an item is based on a proprietary design, or if we can do or make it better, faster or more cost effectively than an outside source, it's a core technology and likely a poor candidate for outsourcing.

Third-Party Logistics arrangements can greatly aid receiving, warehousing, kitting and materials-packaging/presentation challenges, but the same risks and cautions applied to outsourcing also apply to 3PLs (see later discussion).

Global Sourcing is an effective tool for leveraging buying power and for sourcing from low cost providers, *provided* the total cost calculation includes transit time ownership, all port costs, duties, inland freight, additional insurances and any needed adjustments to safety stock levels.

Supplier Consignments improve *apparent* inventory turns by leaving ownership of material with the supplier until we use it. Considerations include:
- Costs inevitably and eventually get built into unit pricing.
- You may still incur warehousing, insurance and management expenses.
- Material may still occupy valuable production space prior to use
- Quality inspections may not be possible until time of ownership transfer (bonded areas)

Additional Software Tools are available and legitimate, BUT, *should not host information or data not included or provided back to the parent MRP database*. *Neither should such tools fail to honor or use standard procurement or planning parameters (such as agreed supplier lead times, minimum order quantities, etc)*. With this caution in mind, several "MRP adjunct" software packages are available and can dramatically aid inventory management efforts.

One such "analysis-level" tool is the "Inventory Quality Ratio" or "IQR®" package, offered directly by IQR® International, and currently used at hundreds of globally-dispersed manufacturing sites. This software provides potential shortage alerts, AND identifies dormant/surplus reduction opportunities down to *part-number or "SKU" level of detail*. Examples of this software's screens and reports are available in the guidebook appendix.

Finished Goods Inventory Management Fundamentals

The purpose of this section is to present recognized finished-goods management best practices for careful consideration and appropriate incorporation to business unit plans and strategies. The process for *planning and scheduling* finished goods production is described in "Production Inventory Management Fundamentals", so this section will focus on the underlying strategies, policies, and principles specific to finished goods inventory management.

These guidelines are organized into (7) categories:

1. "**Whether**" to plan and hold FG inventory
2. "**When**" to build it (or avoid building it)
3. "**What**" to build (and avoid building for stock)
4. "**How**" to plan and manage the program
5. "**Who**" needs to "own" and manage it (and the process)
6. "**Where**" to store and hold the stock once built
7. "**How Much**": Key management metrics, and...
8. "**Toolkit**": Notes on some useful software and other tools

Whether? (to build finished goods for stock)

Finished Goods are typically the largest part of total inventory and working capital in most business units. Some are necessary to meet customer lead time and delivery expectations. Some serve to level production demands over time, best utilize factory capacity, and/or keep within the limits of supply chain support. All should be based on careful consideration and regular review of true costs and benefits.

Stocking of finished goods should be reconsidered at least annually using a total re-justification or "zero-bas" approach, and should address the questions of the true, incremental revenue supported, true total costs incurred, and alternatives (e.g. rather than hold inventory simply because competitors do, could factory lead time be made quick enough for production to replace on-hand FG stocks?). This will help to avoid making or perpetuating the investment simply because of historic or competitors' precedent.

Some recognized principles and guiding concepts include:

- Keep FG stocks as small as possible to minimize their tendency to disguise demand shifts, hide or delay the discovery of quality problems, and increase the risks of damage or obsolescence.
- Work aggressively to compress factory lead time enough that customer lead time expectations can be met from production rather than from stock ("Lean Manufacturing"... possibly including the next concept)...
- Hold unassembled components (WIP) for fast assembly instead of holding finished goods (no labor & burden added until just before sale). This is

particularly applicable to product models with high-variety but low volume, or to models offered with options that cannot be easily added post-production or as field-installed accessories.

- Build to order if customer lead time expectations or requested shipping date permit (unless factory capacity is strained and stock items are available)
- Contain or limit the quantity of stock model inventories using a seasonally re-sized FG kanban system
- Communicate inventory availabilities clearly and regularly throughout the distribution chain, and promote (reward) "inventory engineering" with the sales force and agents.

When? (timing of building for stock)

If building for stock has been decided and defined, capacity utilization, factory productivity, and production load-leveling (heijunka) can all be helped by building the stock product in schedule openings not needed for firm customer order fulfillment.

At the same time, some cautions are in order:

- Understand the risks and costs of a change to customer requirements or market trends when building early.
- Build for realistic forecast or known demand. Simply absorbing capacity can result in unwanted over-production (waste #1), and waste never helps the "bottom line".
- Honor the factory frozen-schedule period when filling production slots. Even if materials are "apparently" available, consuming them may upset Planner/Buyers' allocations and cause shortages later.

Similarly, seasonal pre-building can level peak-season / slow-season production and supply-chain demand, but practical considerations include:

- Review the plan carefully and often. The original business plan and demand forecast may be months old by the time production is actually scheduled, and market demands, customer plans or competitive tactics may have changed.
- Restrict any pre-build to higher-volume, certain-sale models and configurations.
- Avoid using "overtime" labor to do pre-building.
- Communicate plan levels and changes clearly and often with the entire supply chain to ensure their support.

What? (FG inventory model or SKU choices)

Product offering-and-options complexity (and proliferation) complicate FG inventory planning enormously. Good management practice includes at least an annual review of offered models and options *value*, followed by ruthless "pruning". If a small number of customers want obscure or super-ceded models, consider doing only limited builds per firm orders for immediate shipment (no stocking program).

As previously noted, holding popular, higher-volume models is less risky than trying to hold everything. Save and use peak season production capacity for the "unexpected and/or unpredictable".

Develop and enforce ownership policies with regard to large or tailored-product customer orders:

- Include immediate shipment in the terms of sale for low-volume or super-ceded models.
- Avoid scheduling low-volume, stock units if or when it means high-volume SKU stocks will need to be increased during the production period.
- Know, publish and *recover* the cost of large or custom-specification changes or cancellations once procurement and production has begun.
- Know, publish and (wherever possible) *recover* the storage costs of large or custom-specification orders which are ready, but on customer-hold after the requested shipping date.

Design new products for customizing or options-installation on delivery or as easy pre-shipment tailoring & test additions. This would permit the stocking of readily-customized base units, AND greatly ease the forecasting challenge.

How? (the management process)

The S&OP (Sales & Operations Planning) plus production scheduling process is detailed and explained in "Production Inventory Management Fundamentals", but it is key to realize the impact of that process on FG inventory levels.

Understanding and working within defined factory and supply chain limitations is essential. FG inventory created during open-capacity periods can avoid lost orders during full-capacity periods, but major, last-minute changes to production demands are almost always either missed or become the source of other problems.

Regular, reality-based forecast updates are also essential. Original business plan forecasts age quickly and are often defined using revenue units of measure. Constant monitoring and translation to FG unit quantities provides operations with meaningful planning information, and helps avoid building stocks of the wrong product. Accuracy can be improved by forecasting at high or "product-family" level and using Planning-Bills to predict min/max quantities of models within the family. Accuracy is also improved by tracking it and taking corrective action on the root cause analysis of any major "misses".

FG Inventory needs can be minimized by scheduling in ways that maximize real-time production capacity and build-to-order capability:

- Short "model intervals" (e.g. AMED or Any-Model-Every-Day) creates order-promising flexibility and can help level supply-chain loads, *BUT*... production line "changeovers" take time and cost capacity, so finding the right balance is important. For example, if major customers only want or expect weekly shipments (full trucks, lowest freight cost), AME<u>W</u> (any-model-every-<u>week</u>) scheduling would meet their needs and avoid capacity losses from needless changeovers. (Of course, efforts to shrink changeover times will support both flex AND capacity!)
- Where possible, FG inventories should be planned to avoid the need and temptation to force un-forecasted ("emergency") orders into the factory's frozen-schedule period, since these inevitably run afoul of (or create) material shortages and delays to other customer commitments. In hard economic times, *any* customer order is vital, but *not* if accepting it jeapordizes others already promised.
- Care is needed to make certain that scheduling is "true-bottleneck" focused, and that inventory plans are based on true, overall capacity limits. An assembly-line capacity of 1000 units, and any FG plan based on it, is "academic" if the supporting coil-shop or other sub-assembly operations are limited to 900. This means that scheduling-models need to be flexible (bottlenecks get fixed and replaced by others), and FG inventory plans require constant review and adjustment.

A number of software and training tools related to "how" are listed in the appendix of this guidebook.

Who? (process and inventory ownership)

There is no single best answer to the question of "who" should "own" the FG inventory management process or the inventory itself, but some fundamentals apply universally:

- Ownership does need to be clearly defined and singular. Ultimately, "The Corporation" owns all of its inventories, but responsible management and continuous performance improvement (cost/benefit) require that some *individual* has the identified accountability, decision authority and control.
- Whoever has accountability for the FG inventory levels and performance metrics should also be given the influence and authority of Process or Value-Stream Manager, including direct participation in the Business Planning process plus co-leadership (with Sales and Production) of the S&OP process.
- Since ownership will require daily decisions and process involvements, it should be a full-time job for a trained and experienced professional. Assigning it as an "additional responsibility" to a Business Unit GM, Factory

Manager, Sales Director or Warehouse Supervisor is probably unfair to those individuals and to the job.

- Since Sales, Marketing, Production, Finance and other functions all have vested interests in any FG inventory, the FG Inventory & Process-Mgmt Manager or Director will be most objective and effective if the position is at peer level with those functions (and not subordinate to any of them).

Where? (stock location)

Housing FG inventory at the factory where it was assembled can minimize handling, *but* occupies precious production space (or ends up out-of-doors), complicates truck-dock availabilities & management, and usually has to compete with Production for people and other resources. If the inventory is limited and small, these difficulties can generally be overcome.

For larger inventories, factory housing is more of a problem, but a strategy is then needed whether to maintain a large, central distribution center or to move the inventory to ("forward") regional holding points closer to eventual customers.

Forward stocks support fast delivery to any nearby customers, and may even result in transferred ownership to others (Independent Distributors or Franchisees), but are then less available to other, more distant customers. Multiple warehouses also incur redundant staffing & supervision, and overall facilities size can be larger (more costly) than a single center.

Centralized stocking can minimize the total amount of FG inventory needed because demand becomes concentrated… all orders for an item promote flow and turnover, and demand for any item is more constant and regular. The cost for this, however, is necessarily offering a fast-delivery capability for the "regional" customers.

Most large-inventory management strategies follow the automotive parts model, which is a hybrid of centralized + forward stocking points based on the A-B-C or "Pareto" concept. This builds on the observation that ~10% of FG SKU's (Stock Keeping Units) experience regular demand and account for ~70% (or more) of revenues (Type-A), that the next ~20% have only occasional demand and represent ~20% of revenues (Type-B), while the remaining ~70% have rare demand and represent only ~10% or less of revenues (Type-C). Type-A and lower unit-cost Type-B SKU's are stocked at all forward points (local dealerships), while the remainder of Type-B's and Type-C's are held centrally and shipped where needed via priority methods. This satisfies 99%+ of final customer needs, and the premium transportation costs are usually lower than if trying to keep all items at all locations.

Regardless of the model adopted, total inventories can be best managed and minimized by open and accurate tracking and communication of item availabilities across the network… regardless of who technically owns them.

How Much? (Metrics)

The best-managed FG inventories exhibit minimized value (relative to sales) and maximized turn-over, but these are both "rear-view-mirror" metrics. They tell us what happened *after*-the-fact, and do not tell us if we are moving forward safely, relative to lane boundaries and other "traffic". This means that good management also requires real-time *process* measures and pro-active "course-correction" capability. Examples of these include:

- Forecast Accuracy Levels & Trends (input to continuous improvement)
- Aged On-hand Inventory Values (30, 60, 90-days, older?)
- Slow-moving (SM), Non-moving (NM) and "Un-mortgaged" Value$ (and SKU #'s)
    - SM is typically defined as having no visible forward requirements, and no demand for the latest 6 months
    - NM is typically defined as having no visible forward requirements, and no demand for 12 or more months
    - "Un-mortgaged" is on-hand stock without matching, forward demand or orders (unsold)
    - Note: Replacement components are generally not considered NM until demand is nil for at least 24 months… and some stocking may be dictated by warrantee or service-support agreements with customers.
- Shipment values ready per original CRSD but on customer-hold
- Shipment values ready but on consolidation-hold (waiting for all ordered items to become available)
- Shipment values ready but on transport hold (waiting for hauler or vessel assignment)
- Orders (values) shipped more than 1 week late *or early* per CRSD (either gap is a problem for customers)
- % weekly unit shipments met from same-week production, by family or production line (how truly important is the inventory to meeting CRSD?)
- *Note: "CRSD" = Customer Requested Shipping Date (may differ from Promised Shipping Date)*

Helpful Tools

Finished Goods Planning and Inventory Management is an increasingly vital role in most organizations, the supporting technology is becoming advanced and complex, and trained, experienced, full-time professionals in the field are quickly becoming vital members of executive teams. Two very helpful sources of information, education and networking are **APICS** (the Association for Operations Management) and **IBF** (Institute of Business Forecasting). The recent merging of their member-information databases and resources make affiliation with either or both a worthwhile exercise for Inventory Management Professionals.

Computer-Aided-Forecasting tools abound, and can become invaluable when used properly. A frequently used application is "Demand Solutions®", whose choice of stochastic models and algorithms is among the richest, but using them responsibly and fully requires that users have a sound understanding of statistical forecasting methods *and* access to relatively long periods of historic demand data.

There are also a number of "**DRP**" (Distribution Requirements Planning) software tools available. These are the distribution counterpart to manufacturing's "MRP", and can provide a variety of coordinated support functions, including:

- Forecasting
- Replenishment System Planning
- Safety stock sizing and management
- Coordination of Actual plus forecasted demand with product source lead times, on-hand stocks, material in transit and already on-order product to determine new-order needs.

Similarly, there is a wide assortment of available "**WMS**" or Warehouse-Management Systems. These track material arrivals, departures, in-house locations and quantities, personnel and logistical needs going forward, and are invariably worth their cost if used to their design potential.

The best DRP and WMS systems are those that can be linked to the business unit's order entry/management system and the factory ERP/MPS/MRP system. This aids real-time data sharing and eliminates data-entry errors or omissions.

Organizations using "**IQR®**" (Inventory Quality Ratio) Software to help manage raw material and WIP inventories have also found that it can help with FG inventory challenges, such as identifying slow-moving, non-moving and un-mortgaged FG down to SKU level of detail (without time-burden on the main MRP or DRP system).

ILSD (Integrated Lean Systems Deployment llc) can also provide helpful training materials, "kaizen" event support and project consulting services (e.g. "Rate-Based-Planning", "Planning-Bill (PBOM) Development", "Lean Warehousing Kaizen" and other modules & training). Visit www.ILSDllc.com for more information.

So, Who Is Running The Show?

There's an old joke that "A camel is a horse designed by a committee", and the idea remains valid when considering inventory management. Certainly, there are many vested interests, and all have valuable information and input for the process. Certainly, high performance and continuous improvement need to be team-based if they will last.

Just as certainly, however, some*one* (singular) needs to have overall responsibility and accountability. "Shared" overall accountability typically results in "elbowing for credit" when things go well, and "empty rooms" when they don't go well..

The bigger the enterprise, the more likely there will be confused inventory control, responsibility and accountability "fuzziness":

- **Customers** set lead time or requested shipping interval goals, which translate to minimum product stocking levels. They may also expect dedicated or consignment inventories, or the stocking of customized, non-generic products.

- While the factories may physically hold finished product inventories, these (and forward production schedules) may be effectively dictated by **Marketing**. This is not irrational, BUT (having worked in Marketing) the "simplest/easiest" strategy for meeting customer needs (and maximizing sales incentives) is to stock an "endless supply of all models and options". Hmmm....

- **Finance** lobbies for hedge-purchasing of commodities or components with increasing price levels, but often doesn't know or consider the true, total cost of holding the resulting over-stocks. Likewise, clearing obsolete or damaged product and material usually requires a scrap authorization from Finance, who often procrastinates or postpones these until they have a "quarter with excess profits" (i.e., do not hold your breath while waiting).

- **Product Service** expects the stocking of potentially needed repair components from 1st product introduction, even though most will not be ordered for months or even years. They may also have a 10-year or 20-year customer support policy, *AFTER* the product is discontinued. Finally, the "spare-parts" view of "obsolete/non-moving" may be more like 3-5 years rather than the 1-year typical for finished product.

- **Operations** has the constant challenge of keeping production lines running (no "ASWO's") while keeping material moving and in-process inventories minimized... usually in an environment where forecasts are terribly wrong (by an unknown an variable amount), and schedules are changed hourly or daily by "others".

- **Purchasing** is constantly trying to juggle and balance long supplier lead times, inaccurate MRP supplier lead times, excessive minimum order quantity supplier policies, Vendor-Managed Inventories (VMI's), Kanban system sizes, "Global Purchasing & Sourcing" mandates, forward-price hedging needs, "3PL" capacities and capabilities, high Buyer/Planner turnover rates, etc.

- **General Management** has to balance ALL the needs and issues of day-to-day business oversight. Inventory levels are just one of these, so apparently arbitrary decisions and policies are inevitable without a singular, overall Inventory-Manager who has the time and expertise to sort out all the divergent interests appropriately.

The "Inventory Manager's" Role and Qualifications

The "OIM's" or Overall Inventory Manager's title is of no importance. What is CRITICAL is that he or she hold "rank and status" at least equal to the heads of the other vested interests on the SOP... Marketing, Scheduling, Finance, Operations, Purchasing, Service, etc.

The "OIM's" role should be to preside over the SOP meetings, work between meetings with participants to resolve concerns, generally represent General Management in material and inventory issues, select and monitor inventory-related performance metrics, set overall inventory strategies, AND preside over the day-to-day, detail-level "blocking and tackling" needed to avoid ASWO's and enable flow & turns. In summary, the "OIM" has the authority to equal his or her accountability for inventory, and to "manage the white spaces in the organization chart".

"Generalists" tend to learn less and less, about more and more, until they eventually understand little or nothing about everything. By contrast, "Specialists" tend to learn more and more, about less and less, until the eventually understand everything about almost nothing. What's needed is an "OIM" who is neither. Rather, he or she needs to be a multi-disciplinary specialist... with in-depth knowledge about many different things.

Getting good at the job can take years of working in all the key disciplines, and the incumbent should be afforded appropriately high respect, status and compensation.
Accordingly, the "OIM's" initial discipline and credentials do not really matter. Even "Engineers" can make the grade (provided they develop a noticeable sense of humor).

It is a late-career position for high-performance individuals... NOT a "developmental" posting for new-hires or interns. Neither should it be a "second role" for any of the vested-interest Discipline-Managers (not enough time & too many parochial agendas).

Active participation in, and multiple certifications by, organizations such as APICS would certainly be appropriate and helpful, and these should be both subsidized and rewarded by the organization.

The Case for a "Value Stream Manager"

Businesses who truly approach "Lean" status (no waste, full flow, pervasive pull, supplier-distribution integration, standard work and ongoing improvement), such as Toyota, develop and "celebrate" a unique position called "Value Stream Manager" or "VSM".

This person's role is the overall understanding, control and continuous improvement of *EVERYTHING* related to delivering a specific product or product family to its targeted customers... Design Engineering, Marketing, Order-Entry, Planning, Scheduling, Procurement, Supply-Chain, Operations, Distribution, Material, Inventory, Metrics, and (last but not least)... Profit and Loss! They work with everybody, but answer only to the most senior management.

If this sounds like an expanded version of the "OIM", it is, and an "OIM" may not be needed where a competent "VSM" is in place and supported by senior management.

Sources and Fixes for ASWO's

Becoming "Lean" is not the problem! Material shortages have been a part of manufacturing and distribution "forever", and the traditional "when all else fails" response has been to bury the root-cause problems in more inventory. In transitioning to "Lean", non-value-adding inventory is recognized as a waste, understood as an indicator of blocked flow, and (with hard work) systematically reduced.

Not surprisingly, this reduction of cover-ups re-exposes the problems we failed to solve originally, and we finally need to revisit our "operational" issues and fix them:

Common Causes of "ASWOs"

Mat'ls get used before planned
>   Defects
>   Emergency Orders
>   Unauthorized batching
>   Schedule advances
>   Supplier or 3PL system req'ts too low

Incorrect MRP/Planning Data
>   Wrong Lead Time
>   Wrong On-hand qty.
>   Delayed receiving data
>   Short shipment
>   Bill-of-Material error
>   Engr. Change not in system

Material Location Problems
>   Can't find / misplaced
>   Mixed parts delay
>   No assigned location

Late Delivery
>   Supplier late
>   Fab or sub-ass'y late
>   Transporter delay
>   Incorrect ship-by date on PO
>   On-hold by Receiving Inspection

Late PO or Release
>   Poor visual status
>   No / late info to whse or buyer
>   Purchasing backlog
>   Buyer out / no backup
>   Oversight / rare transaction

Mislabeled/Unlabeled Items
>   Supplier / Fabr. Error
>   Receiving error
>   Mat'l Handling Error
>   Lost/missing label
>   Mixed parts in location
>   Label unreadable

Some root causes (and cures) should be pretty obvious, but others may not. Accordingly, it is useful to examine this list more closely.

Materials or Stocks Get Used Before Planned

In the following comments, "3PL" will mean "Third Party Logistics Provider"... a materials warehousing and "kitting" services supplier.

Defects and damaged items can hopefully be replaced from safety stocks (Oops... the politically correct term is now "variance buffers"!). However, "Lean" calls for "buffer" reductions (flow), and "buffers" may not get replenished (especially if "messengers get shot").

The cures include working with suppliers, 3PLs, and fabrication shops to eliminate defects, redesigning packaging/carts/processes to eliminate damage, and rewarding messengers instead of shooting them.

Initially, suppliers may need to do 100% inspection. Life is tough! If they are small, we do need to help ourselves by helping them... even if it means we end up indirectly helping others, including competitors. Large or small, we also need to make certain we have clearly defined "CTQ" (Critical-To-Quality) characteristics AND how we will measure them. It's quite amazing how many specs and drawings totally lack this info!

"Emergency" Orders seem to be an ever-present event. Especially in times of depressed sales, every potential order is treated as "golden". Unfortunately, if the job hits the schedule inside the "frozen schedule period", materials are frequently unavailable and lead times too short for suppliers and/or 3PLs to help. Even where the "emergency" order is satisfied, it is often at the expense of an earlier, scheduled order (making ASWO cause analysis *really* difficult).

The cures include:
- Absolutely blocking schedule changes inside the frozen schedule period
- Increasing finished-product "buffers" to better handle "emergencies"
- Leaving a "space" in schedules for "emergencies", and increasing materials "buffers" to support those spaces.
- Reducing process time so that "lead time" (required frozen period) approaches nil.

Schedule Advances are similar. Depressed sales result in under-loaded schedules, so future orders are brought forward to keep people and processes working. Unfortunately, the decision may be too last-minute for Buyer/Planners, 3PLs and Suppliers to respond with materials.

The cures are also similar to those for "emergency" orders, AND, any chronic schedule advancing has to consider the unfortunate but real possibilities of reducing capacity (working hours or headcount). If depressed sales are "temporary", organizations implementing "Lean" use any extra people-power for accelerating 5S, Waste Reduction, Layout Improvements, Process Time-

Balancing, Setup Reduction Projects, etc... rather than producing product that isn't needed (yet). Chronic scheduling advances is just "mortgaging the future"... The time will come when future orders are all used up, and the only alternative is to cease operations entirely.

Unauthorized Batching is when work-group foremen or team leaders, with all the best of intentions, combine separate work orders... 5 daily orders for 5 pieces of item ABC becomes one order for 25 pieces, Unfortunately, this results in premature use of "20" pieces, upsets the Materials Planner's plan, and taxes suppliers or 3PLs. It can also use capacity for other, intended work, and delay it.

Like hitting your head on the wall, the cure is to stop doing it. Where reason and education fail, I've known organizations to stop releasing more than a single day of work orders, making illicit batching impossible

Service Parts are drawn from manufacturing stocks. This is understandably tempting, but calls for increased "variance buffers", requires *absolute* compliance with (manual) MRP available-quantity balance update rules, AND, Service-Parts order-quantity limits *MUST* be imposed (to avoid leaving manufacturing crippled). Even though it creates (wasteful?) inventory, Service Parts should maintain their own stocks.

Interestingly, the reverse is sometimes a problem... Manufacturing draws from Service Parts stocks (to meet an emergency or replace defects/damage?), and the same cures/rules/limitations need to apply.

Satisfying customers has top priority, but *NOT* if solving one problem just creates another!

Supplier or 3PL requirements are too low... somehow, the "Word" doesn't travel (especially regarding last-minute changes). This is particularly common with info-systems mis-matches that result in delays:

- Production defects or damage goes unreported altogether, or gets "batched" into a "weekly" report, rather than daily or hourly.
- Data transfer requires "translation" or re-formatting during transfer (delay)
- "Negative Available Balances" get arbitrarily "zeroed", so when arrivals finally get posted, the MRP thinks it has more than it actually does.
- Suppliers or 3PLs aren't clear about available-balance and in-transit at any level in the process, and "assumptions become 'ass-u-me' events)

Obviously (but not easily?), the cure is clear, timely data-sharing, and eliminating situations that contribute to data-sharing delays.

Incorrect MRP-System and/or Planning Data

Whether a computer is electronic or "between the ears", the classic first rule applies: "Garbage In = Garbage Out!".

Lead-Times. If a supplier's lead time is too short in the system (or ignored), shortages are almost guaranteed. In fact, most MRP systems contain hundreds of supplier-item lead times set at *zero days!* Trust me and run a report... then cry, then fix it.

What we also forget is that 3PLs and back-shop fabrication or sub-assembly cells also have lead times which will result in shortages if wrong or ignored.

The cures are almost obvious, but not entirely:

- Find and eliminate any "zero-days" lead time settings. If you don't know the value because of a new part or new supplier, replace zero with a conservatively high guess until you do know.
- Less obvious, is to find and fix any response lead time settings less than 3 days. Even if a shop is "in-house" or a supplier/3PL is "next-door", 3 days is about the practical minimum and reliable setting value... ordering, scheduling, producing or pulling from stock, packaging or consolidating, transporting, receiving, and updating of available quantities takes *at least* 3 days, folks... *even if they justifiably call you "Lightning".* A planning lead time includes order-to-*acknowleged* availability, and everything in-between... *NOT* just the production cell or supplier's nominal "turnaround cycle".
- Finally, and in the words of a great US President, "Trust but verify!". The MRP knows when material is ordered and when it is received, and the difference (over 3+ cycles) is the minimum lead time setting that should be used (add at least a day for ordering and BOH updating). Quoted lead times are a good starting point, but the authors may not understand the use of the data, or know what to add for un-owned total process elements. This sounds like a "no-brainer", but almost nobody does it! Too difficult and time-consuming? Really? Compared to an ASWO?
- Whether gathering data from the MRP or directly from product providers, put a reliable system in place to keep it regularly updated! Lead times vary by season and suppliers are also engaged in continuous improvement!

Incorrect BOH (Balance-On-Hand) or Net-Available Quantities

Garbage-In = Garbage-Out!

The causes for incorrect BOH are almost endless, but include:

- Incorrect physical stock-counts (packaging, units-of-measure, estimating, errors)
- Confused "cycle-counts" (similar, but more frequent because the "counters" may be untrained or un-motivated)
- Substitutions! We do it all the time to keep customers happy, but if we "forget" to tell the MRP, we skew the BOH of *TWO* items.
- Mindless MRP BOH adjustments... if posting of an arrival gets delayed in Receiving (Monday and Friday are typically a "fire-drill"), material can get used, and "debited" sooner, resulting in an apparent "negative BOH". Left alone, it will self-correct when postings are completed, but if arbitrarily "zeroed", incorrect BOH will plague the item until accurately cycle-counted (and the apparent but virtual "shrinkage" rationalized).
- Partial or Incomplete Builds... A work or assembly order is started but not finished due to an item shortage. This means the BOH for items that *were* consumed are wrong until somebody tells the computer what is/has/will happen.

The cures get complicated:

- Accurate stock-taking and cycle-counting requires a trained, ongoing team... *NOT* a convention of summer interns or "warm-bodies" from a labor-pool.
- Unscheduled floor actions of *ALL* kinds... substitutions, defects, damage, moves... MUST be transacted on the system on a *TIMELY* (same-shift) basis.
- Event transaction times must be kept equally timely (e.g. Receiving) either via additional hands on overly busy days, or load-leveling to eliminate overly busy days.
- Any manual adjustment to calculated system parameters *MUST* require a written and *Inventory-Manager-approved* rationale.
- A "partial builds" policy must be established and enforced. Ideally, "partials" should be disallowed... rescheduled if shortages are known in advance (and they should be). Where "partials" *are* allowed or inevitable, the product should be "back-flushed" through the MRP (material availability debited per the bill of materials), but missing-item availabilities corrected until resolved, and then re-debited for the corrective use.

Delayed Receiving (or unscheduled floor-action) postings

These are covered under "bad BOH Values" because that's the result. Eliminating delays and creating transaction discipline usually solves the problems, but can prove challenging. Monitoring the type and frequency of floor transactions is a powerful but much under-used method of tracking reporter discipline, and making reporting relatively painless is critical... available data entry terminals, fast forms, clear "why" training, and "immunity for the messengers" (no "3rd degrees or shootings").

Short-Shipments   It *can* happen that suppliers, 3PLs or back-shops can inadvertently deliver fewer items than ordered or invoiced and listed on transfer paperwork. The solutions include:

- Mistake-proofing at the source (actual counts, and defect/damage reports)
- Conscientious counting (not estimating) in Receiving
- Open & honest damage reporting by shippers and handlers (via forgiveness of the occasional accident, and mistake-proofing of packaging or processes.)

Bill-Of-Materials (BOM) Errors

Materials & resource accounting systems (MRPs) rely on scheduled-item bills-of-material (BOMs) to keep item or component availabilities correct. Production or release of a scheduled item results in parallel "ordering and consumption" of all the item's component parts, as listed in the BOM.

Obviously, if the BOM does not list a needed part, or if the needed quantity of a listed part is too small, an ASWO will eventually result.

The cure is a high priority in Engineering for BOM accuracy (NOT a part-time or time-available job), AND a reliable, enforced process for handling **Engineering changes** (material types or quantities) on the floor until the system BOM and drawings are completely updated (manual transaction of materials used.) Here a tracking system for uncompleted Engineering changes and related materials reporting becomes critical.

Also, any change requires an "impact assessment and policy" regarding existing stocks of original parts (use up or scrap), *AND*, a reliable "Revision Control" system that keeps suppliers current & ensures lastest-rev compliance on future shipments.

Material Location Issues

Can't Find, Misplaced or "NIL" Items (Not-in- Location).

While this frequently happens for (occasionally) understandable reasons, the working outcome is an ASWO... *PLUS*, BOH confusion, *PLUS* unnecessary emergency replacement orders with suppliers, *PLUS* emergency freight charges, *PLUS* usually "painful" chats with the Boss, etc.

In other words, avoidance is highly desirable.

Can't Find, Misplaced or "NIL" Items (Not-in-Assigned-Location).

There needs to be a designated place for everything, and everything kept in its designated place. This sound simple, but "stuff" happens, and needs to be both expected and dealt with aggressively.

"Locations" can be too large or vague to make finding items reliable ("over yonder" or "on the line"). I even know one site where ~75% of material is located in "Location O-D" (Out-Dere or in-Production... someplace). They can also be so overly-precise and complicated that material handlers need a university education to find anything. A good compromise is a system that puts any specified item within "easy sight" when the handler gets there.

Spaces can be full when a handler tries to place new quantities, so he/she substitutes the nearest open location. Make updating the computer mandatory, easy and fast!

Mixing parts in a single location or container is disastrous, but happens frequently, especially in rack-storage areas or among low-usage items. If several parts must share a location, each should have a separate container and mistake-proof labeling. Be careful, however... Avoid "burying" one item with another, requiring "triple moves" to get to the buried item or product (move the unneeded item, get the needed one, replace the other).

High-rack, top-shelf items are difficult to find and to accurately identify, so use low/floor stores and locations wherever possible, and make high-stored item labels "super-sized".

Bar-coding all materials and locations is also recommended. It reduces human error and eases action-reporting.

Late Deliveries (Suppliers or 3PLs)

It happens, and nobody is perfect, but apparently late deliveries by a supplier are a classic instance for using the "Çinco Porqués" or "5-Whys" problem-analysis tool.

Overly short or ignored provider lead-times (in the planning system) have already been mentioned. Suppliers need to continuously improve lead times and hold reasonable "variation buffers", but in the meantime, their lead times are what they are, and ignoring them is just plain stupid. I came across one situation where all Purchase Orders had to be sent via fax, and 22 Buyer/Planners had to share a single fax machine. This predictably added 2-3 days to the PO transmittal time, and 2-3 days to normal deliveries: a self-inflicted and mindless situation.

Truly Late Sources Suppliers, fabrication shops, shared-resources (like painting), sub-assembly cells and 3PLs are just like us... *FALLIBLE!* Occasional failures to deliver on time can have the same, excusable causes that make us late with our customers. Chronic failures are a "call for help". Switching sources is only a consideration *AFTER* making sure we are using and honoring correct lead time and info-transfer elements, and AFTER making certain we don't confuse actual receipt and transaction-posting dates at Receiving.

Transporter or Transportation Delays Remember that transporters and the transportation process can introduce delays separate from the suppliers themselves. Shipments can be "on time", but arrive "late" due to delayed sailings or flights, port delays or strikes, customs confusion, or delivery-vehicle breakdowns.

Where infrequent, a modest "variance buffer" can resolve the potential ASWO. Where chronic, we need to increase planning lead times and/or switch to more reliable transporters and processes. Minimizing "Cross-Dockers" can reduce time and damage. An "organized" customs broker can electronically submit advance bills-of-lading to expedite customs clearance. Truckers can be tracked for performance and prioritized for contracts. Ultimately, long-distance suppliers may have to be replaced with "locals", if production and ASWOs prove unmanageable.

Requested Dates... Avoid confusion between requested ship-by and deliver-by dates on POs and Releases. Any requested date should be clearly identified as a ship or receive-by type (many are not). The choice depends on who owns or contracts the shipping. If the supplier owns or contracts the shipping, a deliver-by request makes most sense. If we own or contract the shipping, a ship-by request is more appropriate, and it should allow for (pessimistic?) transport time. The point is to be clear and logical. Above all, be specific! "Needed in fiscal week 37" invites disappointment.

Receiving Delays... It is not unknown for an apparent delay or lateness to actually be items on "in-house quality hold" or "in-house documents-validation hold". Here, cures include enforced turnaround time standards for in-house procedures, and document/PO validation system improvements as needed. Time is money.

Late Purchase Orders (POs) or Contract Order Releases

If a supplier does not get an order in a timely fashion, he or she has little to no chance of meeting delivery needs. Unfortunately, this sometimes happens, AND it gets recorded as a "late delivery", making ASWO cause analysis difficult. Accordingly, any ASWO investigation should include the PO or release transmittal date for comparison to Receiving date and to nominal supplier lead time.

But, what caused late PO or Release transmittals?

Earlier, we heard of the case where multiple buyers had to share a single fax machine (instead of using the e-fax capability of their individual PCs). There, late advice to Suppliers was clearly understandable, but there are a number of other causes to also consider.

Poor Visual Status   The MRP may recommend a buy, but if that is not acted on immediately, and the list can get buried and delayed on a Buyer's desk. Similarly, if kanban discipline is poor (e.g. working from 2 or more containers at the same time), those signals can also be delayed. Either way, timing-discipline is an item for education and performance feedback. Also, gravity-racks or vertical stacks can discourage working from multiple kanban containers simultaneously, and simply attaching all recommended buy lists to a sheet of large, red construction paper makes them difficult to ignore or misplace.

If On-Hand Balance accuracy is in question, and delaying Buyer decision-making, getting BOH re-calibrated is essential. Meanwhile, a readily available special-counting team needs to be at buyers' disposal.

Late info to the 3PL, Warehouse or Buyer   Last minute schedule changes or MRP updates will result in delayed signals to Buyers, and delayed POs or releases. This is another, compelling reason that last minute schedule changes should be avoided, and why any I/T delays to MRP updates should be aggressively analyzed and eliminated.

Purchasing Team Backlog   Is the Buyer team overloaded? If a PO or Release takes 15 minutes to review and activate, are buyers expected to look at more than 4 situations per hour (30-35 per day). If the MRP is only fully updated once per week, is that resulting in a "mountain" of purchase recommendations that takes 3-4 days to clear? And again, is all the needed decision-making info readily available and trustworthy, or must Buyers make counting trips to the floor or warehouse to verify true BOH? Is automation and e-commerce being fully utilized to minimize ordering time and errors?

Buyer Absent... Without Backup. Buyers are people, and are occasionally absent. They and the organization also benefit from regular training. Is there a trained "backup" person available (like the Purchasing Manager)? Are file-organization and other methods standardized so that others can find and deal with the work?

Oversight / Rare Transactions   Long lead time items usually get handled manually, without the help of the MRP, and may be ordered in quantities that result in review of needs only once or twice per year. Not surprisingly, these are easy to forget or overlook, so late POs and deliveries follow. Regardless of transit time or normal packaging, try to limit order quantities to less than 60-days of supply.

Another tactic is to have one buyer specialize in and focus on these "infrequent order" items, and develop a disciplined review process.

Late or Missing BOH Update (back-flush)... Part of most MRP or DRP updates is recalibration of BOH and assessment of new order needs. If this is delayed, or (more commonly) if it takes place without having *ALL* production or shipping reports as inputs, the chances of an informed Buyer decision are small. It has been said before, but is worth repeating... any and all floor/warehouse actions *MUST* be reported or logged to the MRP or DRP as soon as possible, and at least during the same day or shift as the action. System logs can reveal who submits what kinds of information and when, and anybody who is chronically "casual" needs reminders added to the process.

Un-Labeled or Mis-Labeled Product or Material I prefer each ITEM to be clearly labeled, especially if there is any chance for confusion with similar parts, or if there will be any time between removal from its container and use on the product or placement in the shipment. Sadly, suppliers and back-shops prefer to label lots rather than pieces, and to use easily removed labels (which can disappear accidentally during material handling).

In addition, labels get torn, dirty or "smudged", making then illegible, and inadequate or poorly organized storage leads to having multiple or mixed parts in the same location.

A reliable-systems approach suggests that every location and item be bar-coded or Radio-Frequency-Identified. This minimizes confusion, illegibility, label size, and the time & effort needed to apply labeling. Obviously, minor items like screws, nuts & bolts cannot be individually labeled, so other forms of ID and mistake-proofing are needed in the process (e.g. a labeled nut, bolt and and screw sample-board at the point of use).

I encountered one case where parts received one of 6 colors in the paint shop, but ALL forms of that part had the same part number... regardless of paint, no paint, or color! Do you suppose they experienced BOH accuracy issues and ASWOs?

Major or multiple "NEWS"

While not on the list of ASWO sources above, it is observable and true that any major new element in the process can result in some ASWOs. Two major new elements makes ASWOs highly likely, and 3 or more *guarantees* them. "News" can include:

- New product in manufacturing or distribution (with only "forecasted demand")
- New component (requiring 1st item qualification)
- New Supplier (requiring overall qualification, monitoring & training)

- New Buyer/Planner (even if "experienced" elsewhere)
- New I/T staff (re the MRP or DRP system)
- Broad scale "out-sourcing" program
- Broad scale "in-sourcing" program (reverse out-sourcing... popular in recessions)
- Broad scale "global sourcing" program
- A switch to "Lean" tools and methods (e.g. lot sizes, kanbans, planning-bills-of-material, etc

These all lead either to demand and usage uncertainty or to unknown quality and delivery-reliability or lead time performances. The usual "hedge" is to put conservatively large "variance buffers" in place, then suddenly eliminate them (invite ASWOs), or forget to remove them (ugly, excess stock).

Any "major new" should have an assigned and accountable "Transition Manager" in place until the process re-stabilizes.

A Note on Problem Solving or Root-Cause Elimination

A reliable truth is that 85% (or more) of any problem is due to the process rather than the people, and the people who do the work are the best source of mistake-proofing ideas. Even if they are not "optimal", the people will *MAKE* them work out of "ownership".

People inherently WANT to do consistent good work. If they seem not to, the odds are that they need better tools, better information, mistake-proofing aids, or a redesign of the entire process.

Sources and Fixes for Excess or Surplus Stocks

For purposes of this text, "Excess" will mean any over-stock of items that DO get used, while "Surplus" will mean stocks of things that do NOT get used, including any "obsolete material (such as those promo kits from 1988 that Finance is reluctant to scrap).

Here, there is both good news and bad news.

The good news is that excess and surplus most often result from the same *types* of process problems that cause ASWOs. Any time you are working to resolve an issue, you have the opportunity to fix *two* problems... overstocks *AND* ASWOs!

The bad news is that the problem solving will need to consider both extremes, and the solution will need to work both ways.

Here are some examples:

Using material *before* planned can cause ASWOs, while **delaying or postponing the scheduled** use of materials can result in Excess stock. Deciding to operate below full capacity for a time may be prudent, BUT, materials already en route will (e.g. from "Timbuktu") usually keep coming. You may also have firm commitments to take minimum quantities of some supplier items, regardless of usage, because they are working in advance of your schedule to ensure timely deliveries. Customers who delay or cancel their orders move the operations schedule back for you. The "ASWO solution" includes heightened schedule discipline, a frozen schedule period, and a rational approach to handling of emergency orders. Extending this to also be an "Excess solution" might require defining and adopting a "pipeline draining" period before any major schedule cutback, an effective supplier-alert system, and order-cancellation charges for last-minute customer changes or cancellations.

In one situation, customer changes and cancellations were a major materials and scheduling problem. As it turns out, they knew well in advance, but the information was not getting to the factory. In the end, the improvement team concluded that the Sales force was not tracking open order status because they were measured and rewarded for *bookings*. The measures and rewards were accordingly revised to monitor *on-time shipments*, and most of the confusion went away, *without* penalizing and angering the customers.

Incorrect MRP Planning Data can drive Excess as easily as it drives ASWOs. Supplier lead time data that is too long results in premature new orders. If BOH or on-hand quantity data is too low, the MRP triggers unneeded replenishment. Suppliers have been known to mistakenly duplicate shipments, and the Receiving system was not programmed to look for them. Product Bills-of-Material (BOMs) can mistakenly include unnecessary items or quantities. Engineering can make a design change, but it somehow fails to be reflected in the MRP-BOM (or the new item is added, but the old one not removed because "using existing stock is OK).

Here, we can end up with TWO items in Excess status, one of which becomes "obsolete stock".

A variation on the theme is planning parameters that are technically correct, but counter-productive. An example is a minimum-order-quantity or order-multiple quantity that amounts to many days or weeks of supply (automatic Excess).

As with ASWOs, the solution is constant vigilance regarding the planning parameters. Some locations write standard MRP exception-queries to flag troublesome lead times, MOQs, OMQs, safety stocks, etc. Advanced inventory management software such as "IQR®" does the same, AND automatically translates BOH, safeties, open orders, etc into "days of supply" for Buyer/Planner review.

Poor Location System Design creates ASWOs if material cannot be readily found. The Excess consequence comes in the form of an (unneeded) emergency order for more of the missing item. It can also result when material is kept in "4" locations, but the tracking system will only handle "3".

Best use of storage space requires flexible locations and multiple locations for a single item. "Lean Manufacturing" usually finds any single item kept in central stores, area "supermarkets", and (possibly several) points-of-use on the production floor. If a 3PL service is involved, you can add several more locations to the overall system.

Fixing Location-based ASWOs requires careful layout, disciplined handling, and mistake-proofing tools such as bar-coded product and locations. In any large warehousing operation, a full-blown Warehouse-Management-System becomes essential. Fixing location-based Excesses follows the same plan.

Late deliveries cause (temporary) ASWOs, but **Early Deliveries** can cause (temporary?) Excesses. In a distribution center, early arrivals create the additional problem of inadequate space... no time to "make a hole" or consolidate and tidy-up the intended location. The point is that material needs to arrive "just-in-time" to avoid both ASWOs and Excesses, and to "flow". This is yet another reason that accurate lead time data and schedule discipline are so critical.

Similarly, just as late ordering or releasing by Buyers & Schedulers can create ASWOs.

Early or Premature Ordering and Releasing can create Excesses. Inaccurate BOH data, overly long lead time data, excessive MOQs, and uncertainty about quality or delivery reliability aggravate the problem. Unconstrained bulk-buys (price-hedging, transit time, uncertain availabilities, etc) fuel the fire, and then there is always the "reward" system. It is a sad reality that in most operations, there are "severe consequences" for ASWOs, but relatively mild consequences for Excess, and few rewards for minimizing inventory and maximizing turnover (flow).

A rigorous approach to stock level planning is essential… one that promotes "just-in-time", but also gives Buyer/Planners meaningful data about an item's best-choice management method (Micro, kanban, FIFO-lane, Min/Max, VMI, etc.), impact on overall turnover (A-B-C) and allowable weeks of supply (an overall target of 26 turns does _NOT_ mean that every item stock must be kept at or below 2 weeks of supply!).

An equally critical part of any "fix", is ensuring that lead time, transit time and border delays are factored into "global" sourcing mandates, AND that decisions to bulk-buy-early include direction re balancing via reduced stocks on other items. There is always a fixed amount of space and money for materials. Adhering to the "budget" does not happen by accident.

Unlabeled or Mislabeled items create both ASWOs _and_ Excesses. Apparently missing items get needlessly replaced, and wrongly-labeled items sit there forever (folks do know they're wrong)! The fix is closely linked to that for poor location control:
- Clear labeling of all items (not just containers)
- Use of bar-coding or RFID (including by all suppliers)
- Labeling in Receiving (if not done by suppliers)
- Use of smudge-free labels where appropriate
- Adding item photos to labels (so mis-labeling is obvious). Note: this is EASY if the right bar-code labeling system is employed!

New Process Elements can easily cause ASWOs. New products or components are seldom perfectly documented on release. New Suppliers, Buyer-Planners or IT Gurus will definitely go through a learning-cycle, even if highly experienced. A switch to outsourcing, global sourcing or "in-sourcing" is bound to create confusion.

Incorrect or missing information drives the ASWOS, but "fear" drives the Excesses. People know that major changes involve confusion, mistakes and risks, so "Rule 1" (_NEVER_ stop operations for a material shortage) leads to huge "variance buffers" or safety stocks. These buffers may be prudent and necessary during the transition, but are seldom based on facts & data, and almost never eliminated after the change has stabilized (resulting in permanent Excess).

Don't leave these critical decisions to Buyers and Planners who are seldom part of the change planning team. Insist that detailed materials-impact plans be a part of any change-transition process, and that "temporary" buffers are eventually eliminated. (Many sites keep them in a separate location so that they do not become part of the "normal scenery")

Finance seldom causes an ASWO, unless they delay supplier payments or cancel customer credit lines (resulting in last-minute schedule changes). They CAN, however, and unintentionally, drive Surplus accumulation.

Failing to dispose of damaged, defective or obsolete material & product creates "Surplus", but approving a "scrap ticket" or authorization results in a direct reduction of net profits, for which "reserves" (budgets) are often inadequate.

Operations needs to accurately value "Surplus" at annual business planning time, and ensure that adequate "obsolescence and scrap" reserves are in the budget and profit plan. Then, everyone needs to "clear the deck" monthly, so that the "nuisance" does not become an "unmanageable and unaffordable monster".

Unlike fine wine, Surplus material does *NOT* improve with age. Every day it is left in-house it incurs storage, handling and insurance costs, and its salvage or scrap value can shrink.

What about requiring **Supplier Consignments**? Requiring suppliers to hold or own materials until actually used gets them off of the "books". This reduces apparent on-hand values, and appears to boost flow and turnover.

Remember, however, "there is no such thing as a free lunch, and if a thing seems too good to be true, it probably is." Holding inventory incurs cost, and suppliers *WILL* eventually find a way to build consignment costs into their prices and fees. Meantime, you will have lost a lot of control, increased obsolescence risks, complicated quality control efforts, and necessitated an additional (time-consuming) bureaucracy at the point of materials release/transfer.

About the only situations where consignments can really help are when transit time and shipping quantities are outlandish (global sourcing from "Timbuktu"), or when an item has huge value (e.g. jet engines).

Another source of Surplus *AND* Excess inventory is **product proliferation**... Model B replaces Model A, BUT, one or two influential customers still want Model A. This unfortunately means you continue to make and stock Model A, *and* to buy and stock unique components for it, while *also* stocking for and producing Model B. Clearly, it is best not to turn down orders, *but*, Model A should be built on a strict procure/build/ship basis, just once or twice per year, rather than be treated as a current model... no stock (unless the customer wishes to hold his/her own)! This may sound "hard", but Model B was undoubtedly touted by Marketing as "the future", so it is best to move on.

Material Location Systems

Creating and using a good material location system is, in American slang, a "no-brainer"... meaning that it does not require a PhD degree to understand the need and benefits.

If a material handler reports, "Cannot Find" or "Not In (assigned) Location", you know it has just cost the enterprise valuable time and considerable cost to search for something in vain. The second effect is an ASWO or shortage, followed by an emergency replacement order (more cost). The third effect is then a surplus of the item (once the original quantity *is* finally located).

While this is quite obviously a problem, it is amazing how many operations do not have a workable location system, or they have one and do not use or enforce it. This is especially true in smaller operations, where an emphasis on "team effort" is corrupted to embrace "handler's choice" materials organization. This might seem OK, but can quickly lead to the "Missing Man Syndrome" where production comes to a standstill because the original handler is absent. It also assumes that the handlers have advanced training and insight to the needs of retrieval efficiencies, physical stock-taking, cycle counting, safety standards, and existing locations of the same item.

There is no universally "best" location system because every enterprise has a unique assortment of materials, building, people, information systems, partners, regulations, etc. There are, however, at least 5 Design Factors and 8 Guidelines to consider when creating or assessing your system.

Factors in Location System Design

Building (or buildings) size and layout
Material or product size, weight and packaging
Tracking-System capabilities (especially if computer-based)
Partner Communications, Data-Sharing and Methods Alignment
(3PL or Warehouse)
People Safety and Productivity

Building (or buildings) size and layout are a clear consideration. If there are multiple buildings, they should all have similar location systems, and the location codes need to include a building designation. Bigger buildings need a more sophisticated location system than smaller buildings. Older buildings may have "wings, additions, partitions and multiple rooms", all of which need to be identified in the location-naming system.

Material or product size, weight and packaging will strongly influence the choice of rack storage, floor storage, bin storage, labeling, handling needs, staging or kitting space & locations, etc.

Tracking-System capabilities (especially if computer-based) are a major consideration. Simple (usually older) systems may only allow 1 location per item, so those locations need to be large enough for maximum expected quantities. More complex (and modern) tracking systems allow for multiple item locations and first-in-first-out (FIFO) control. They can optimize area/volume utilization, but require more complex location-labeling, data-recording and handler-training.

Partner Communications, Data-Sharing and Methods Alignment? There is an increasing trend, especially among "Lean Aspirants" to enlist outside service providers to help with materials and inventory management. These can be simple warehouse operations or advanced 3PL's (3rd Party Logistics Services), depending on whether the need is simply more space, or a full range of services that might include procurement, receiving, storing, staging, kitting and/or packaging and shipping.

Either way, the location system needs is ideally parallel for both parties, and compatible with both participants' tracking systems. Aside from everybody knowing where things are, this is essential when it comes time for physical stock-taking or periodic cycle-counting.

It is also critical when it comes to managing finished product inventories. Do you suppose that Toyota factories consider their downstream inventory levels before assembling more vehicles?

Regardless of who owns material at any given time... you, the 3PL, a supplier or distributor, its quantity and location is critical information for all.

People Safety and Productivity should *always* drive the location system strategy and details. This means the system must be logical, easily learned, and that using it should be seen as a help rather than as a chore. Handling equipment should be adequate, conform to all safety standards, and be the focus of training for all users. Data gathering equipment (such as bar-code readers & label-makers) need to be equally adequate, easy to understand, and the focus of training.

Most importantly, safety should be paramount. Large, heavy materials or product should be kept on the floor wherever possible. Sharp objects should be packaged and labeled accordingly. If rack storage must be used, the racks should be adequate for the weight imposed, bolted to the floor, equipped with full decking, and labeled in readable format *from the floor*.

Here's a checklist to help guide your design choices:

Location System Guidelines

1) **Simple and Easy to Learn**
2) **Functionally exact (but avoid precision that includes complexity)**
3) **Handler Safety and Productivity Based**
4) **Eases Stock-Taking and Cycle-Counting (speed, accuracy, safety)**
5) **Accommodates appropriate container systems**
6) **(Ideally) allows for multiple item locations**
7) **Meets MRP, WMS or Tracking System limits and coding needs**
8) **Consistent format for locations and contents**

1) **Keep the location system simple and easy to learn.** New materials handlers should be able to master it in a day or less. Create "zones: from obvious features such as buildings, central stores areas, assembly areas, Receiving, Shipping, truck trailer parking lots, 3PL facilities, etc. Assign "street names" (and create "intersections"). Hang large, overhead area labels. Position "you are here" maps at all major aisle intersections. Label everything with consistent-format bar codes or RFID tags (Radio Frequency ID), and make the tags large enough to be read from a floor position (10-meter BC readers ARE available).

If you have ever been to Salt Lake City, you will know what is intended. There, all central streets and avenues are named for the number of "blocks" North, South, East or West of the Mormon Temple Square. Easy.

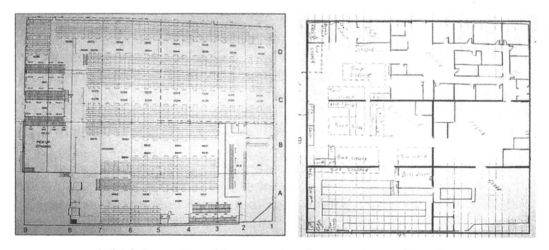

Which Layout would you rather have to learn & use?

2) **Keep the system functionally exact (without creating complexity).** Simple reference to building support columns or locations such as "OD" (meaning "Out Dere") are probably too indefinite. The locations need to be definite enough that the desired item is readily visible from that location point. Minimize rack storage and favor marked floor locations. Where racks are used (*central stores*

only) the location codes need to specify building location, specific rack, rack row and column. Avoid mixing parts in any single location, but do allow for the staging of "kits" with "kit numbers or names" (Kits and lean materials are quite inseparable).

Spot quiz: Which 2 photos depict best space use and people productivity?

Which two are most likely to cause ASWOs or Excesses?

3) **Design around material handler safety and productivity.** Keep spaces for large
and/or heavy items and frequently needed items down low. Insist that all racks have durable decking to prevent drops or spills. Make all item and location labels big enough to allow reading from floor level. Consider integrating item photos to item labels. Make the system capable of 3-way queries… "Where do I find an item?" or "Where can I store an item?" or "What is *supposed* to be in this location?". Make certain that aisles are wide enough to move big items around at floor level (NO fork trucks moving with raised loads!). Where possible, use location systems with picking and put-away routing logic

4) **Consider Stock-Taking and Cycle-Counting Ease.** Both activities are "non-value adding (customer view) but necessary, so designing the location system to support them and waste minimum time and travel makes sense. A-B-C classification is discussed in detail later, but "A" items get counted at least 4

times more often than "C" items, so keeping the "A's" easily accessible will save a great deal of time and effort.

5) **Use simple, appropriate container systems...** meaning "form should fit intended function", generic containers are probably sub-optimal. Limit "tubs" to 3-4 sizes only, and standardize their colors (e.g. blue for components, red for defective items, yellow for sub-assemblies, etc). Use gravity racks for tubs in Supermarket or points of use (allows quantity control and prevents working from multiple tubs). Avoid using "buckhorn" or similar crates and cages (re-useable, but typically only 20-30% full and prone to damaging contents). Ban shipping pallets (and fork lift trucks) from all production areas (dangerous, space wasteful and almost never holding just needed quantities). Design special racks to store long items vertically. Ban production area racks or stacks more than "nose-height", and limit production area quantities to just what is needed in the short-term (3-4 hours max).

This sounds like it could require extensive handling teams, but is actually quite efficient once standardized, AND it keeps most material in legitimate storage areas where they can be tracked and accounted for. Delivering pallets of material to points of use is dangerous, creates over-crowding, wastes space, and results in "lost" (or cannot be located) quantities.

"Buckhorn" cages are polular, and can be re-used, but typically use only 20% of the volume, and, they all look the same!

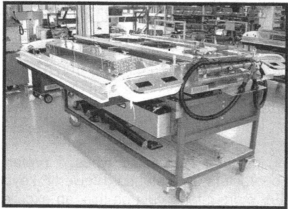

This assembly cart features component drawers underneath, so there is no need for inventory on the assembly line, or handlers to deliver it!

Examples of "too many" locations and labels. Both items should be scrapped and removed, rather than left in "inventory".

6) **(Ideally) allows for multiple item locations.** The best use of storage space (cube) results when the location system allows for multiple location assignments for any part. When one location filled, another is assigned, and the system tracks the "age" of contents so that FIFO usage (First-In-First-Out) is achieved. This also helps in situations where delivery quantities are MUCH larger than can be efficiently stored in production areas (e.g. a container load of items from half-way around the World). The alternative is simpler (fixed location for each item), BUT, the spaces allotted must be big enough to hold the maximum in-house quantity ever anticipated.

7) **Location system is consistent with MRP or WMS (Warehouse Management System) coding formats...** meaning the info-system must accept and be able to decipher the desired location codes, and/or the location codes must be built around the system formatting limitations. Many older MRP systems are very limited with regard to both multiple locations and location coding. If yours is more than 2 generations old, it may be time to update.

8) **Use the same labeling system for both locations and their contents...** so that you can avoid multiple system costs, interface complications, maintenance, training and confusion. Some enterprises are experimenting with RFID tagging (Radio Frequency ID), especially where material or product is constantly moving or unmanned reading is helpful. In most applications however, I prefer a system that is readable by both machines and people... such as a bar-coding system.

These are *NOT* overly expensive (starting under USD 1000), the vendors can provide turn-key, nearly plug-and-play systems and training, *AND* can do likewise with suppliers so that it is unnecessary to apply needed labels in Receiving.

While there are numerous, different bar-coding types, I prefer "Code-128" because it includes all the characters on a standard "qwerty" keyboard, and is accordingly very flexible. For more information on this, an excellent source is the "Bar Code Primer", available in .pdf electronic format from Worth Data Inc. (www.barcodehq.com).

The following are excerpts from a "Supplier Info Package" developed by a well known manufacturing business. Most is self-explanatory, but notice the emphasis on label locations for material, and the idea that materials can be largely put into locations without fear of hiding the labels (The complementary idea is to design locations so that they do not hide content labels!).

In general:
- Minimize the need for keyboard or keypad input (errors approach 15%)
- Standardize location & label designs and placements
- Consider including item photos on labels
- Consider code readers at kanban points of use ("need-more" signals to avoid removable cards?)

Basic, Bar Code Labeling System

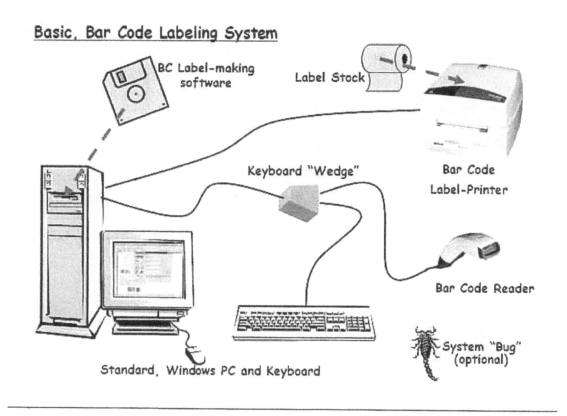

BC Label-making software

Label Stock

Keyboard "Wedge"

Bar Code Label-Printer

Bar Code Reader

System "Bug" (optional)

Standard, Windows PC and Keyboard

Sample Bar Code Label

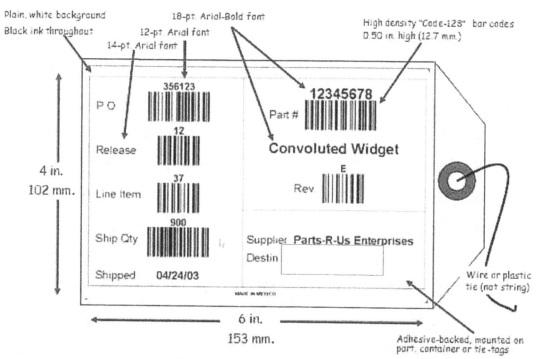

Plain, white background
Black ink throughout

18-pt. Arial-Bold font

12-pt. Arial font

14-pt. Arial font

High density "Code-128" bar codes
0.50 in. high (12.7 mm.)

356123

P O

12345678

Part #

12

Release

Convoluted Widget

37

Line Item

E

Rev.

4 in.
102 mm.

900

Ship Qty

Supplier Parts-R-Us Enterprises

Destin

Shipped   04/24/03

MADE IN MEXICO

Wire or plastic
tie (not string)

6 in.
153 mm.

Adhesive-backed, mounted on
part, container or tie-tags

Note: Advanced BC labelers can even include an item photo!

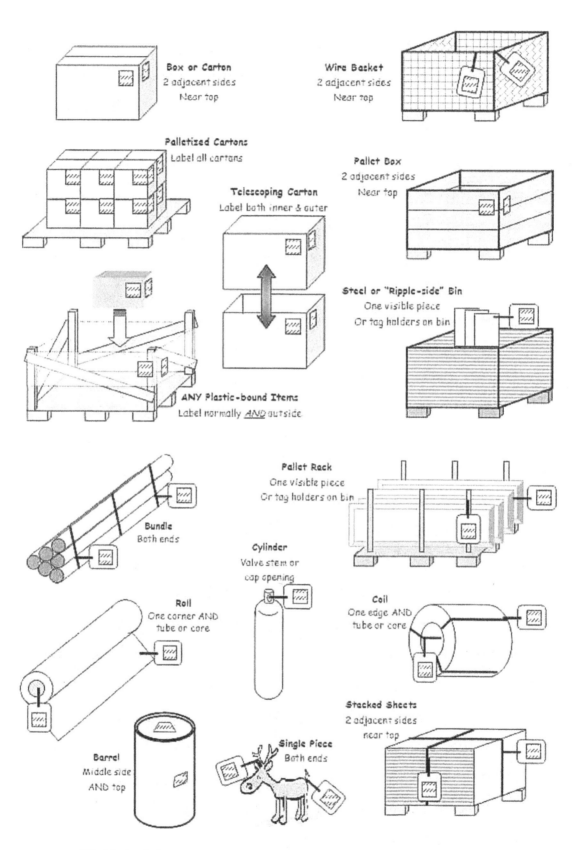

Box or Carton
2 adjacent sides
Near top

Wire Basket
2 adjacent sides
Near top

Palletized Cartons
Label all cartons

Pallet Box
2 adjacent sides
Near top

Telescoping Carton
Label both inner & outer

Steel or "Ripple-side" Bin
One visible piece
Or tag holders on bin

ANY Plastic-bound Items
Label normally _AND_ outside

Pallet Rack
One visible piece
Or tag holders on bin

Bundle
Both ends

Cylinder
Valve stem or
cap opening

Coil
One edge AND
tube or core

Roll
One corner AND
tube or core

Barrel
Middle side
AND top

Single Piece
Both ends

Stacked Sheets
2 adjacent sides
near top

Multiple BC Labels are less likely to get "buried" or hidden

Good & Economic Bar Code Labeling Software

- Respected, reliable source
- Down-loadable from the web
- *Selectable user languages*
- WYSIWYG control
- Full ODBC / Excel 'Wizard'
- Free, advanced drivers for all printer types/brands
- Cost under US$500

A free, demo version of this software is available on www.seagullscientific.com if you want to experiment.

I created this label template in about 20 minutes, following a half-day of self instruction. In use, each label takes only seconds to load and print. Label job data can be entered on the screen, or selected from an Excel, Access or other ODBC-format data file.

Eric

Forecasting (The "F-Word")

Everybody tends to blame incorrect forecasts for *both* ASWOs and over-stocks, perhaps because nobody is quite sure who created and owns the projection.

What puzzles me is why so very few take corrective action, starting with measuring the difference (error?) between actual and forecasted demand.

This is such a big subject, that the "Schmidt-Creek Paddler's Guidebook Series" will ultimately include an entire volume, called "F.A.S.T (Forecasting And Scheduling Techniques)". Meanwhile, here are some preliminary tips on the subject...

The only thing one knows for certain about a forecast is that it *will differ* from the real quantity being estimated. Notice I said "it will differ", not "it will be wrong or in "error". Since the perfect crystal ball or computer model has yet to be found, the real question is, what to do about it?

Some logical strategies include:

- Reduce production and delivery lead times so much that the forecast becomes irrelevant... interesting and worth having, but largely academic.

- Track and quantify the difference between estimated and actual demand! This is the basis and starting point for any rational improvement, and at very least, the organization will learn how much adjustment or "Kentucky windage" can/should be applied to near-future guesses.

- Establish a "collaborative forecasting team and process" as part of the S&OP meetings, where Operations presents an advance statistical projection of history, Marketing & Sales overlay their insights to changing market conditions (competitive actions, new/lost customers etc), Procurement adds supply-chain boundaries, and the Value Chain Czar or General Management leads all parties to consensus. (The statistical projection greatly helps people calibrate their own opinion and guess).

- Stop trying to create and use a single-point forecast (practically impossible to get correct). Rather think and calculate in terms of probable minimum-vrs-maximum values... i.e., in terms of a probable range. Then plan people for the middle value, and materials for the high side. This pretty well eliminates ASWOs without creating huge surpluses, and trades on the observation that organizations can almost always "flex" output if needed materials are available.

- Forecast at highest practical level (product family rather than individual model or variant) and develop PBOMs (Planning Bills of Material) that use historical mix probabilities to convert the high-level projection (min/max range) to lower-level min-max ranges for family sub-categories, models, options, and even individual components.

- Don't build or stock everything to forecast! Some predictable % of customers may not want or need delivery "at-once" or faster than the build-to-order lead time. This sounds obvious and easy, but it is not, and the most difficult aspect can be reshaping the thought processes for Sales and Order Entry. Sales will need to probe for true delivery needs and expectations *before* making potentially and unnecessarily short promises, and Order Entry will need to think each order through *before* simply trying to ship from stock then booking production (for some orders, the "booking production" option will come first!) The other challenge is to data-mine history for customers' delivery expectations and preferences (Most data bases show order-to-shipment, which is what you did vrs. what customers asked for or wanted.

- Don't use suppliers with lead times longer than the forecast interval. This is like the fellow who raised his arms toward the ceiling and said to his Doctor, "This hurts.", to which his Doctor replied, "Then stop doing it!". If you cannot accurately forecast requirements over a supplier's lead time, your choices include predictable ASWO's, predictable surpluses, or plenty of customer disappointment. If you are blessed with a Buyer/Planner who seems clever and reasonably successful at juggling long-lead time situations, don't just celebrate… provide recognition and get them on the S&OP and forecasting teams.

I suggest starting with rational (quantitative) "difference" tracking:

Product Family: Convoluted Widgets

Actual Month	Jan	Feb	Mar	Apr	May	Jun	Jul	Aug	Sep	Oct	Nov	Dec		Avg.	+/- 2σ	95% Confidence Range Fcast - Actual
Actual Demand	100	120	125	135	150	135	135	127								

Forecast Month	Dec	Jan	Feb	Mar	Apr	May	Jun	Jul	Aug	Sep	Oct	Nov				
Forecast Value	95	112	118	125	141	125	125	120	115							
Difference	-5	-8	-7	-10	-9	-10	-10	-7								30-day
% Difference	-5%	-7%	-6%	-7%	-6%	-7%	-7%	-6%						-6%	2%	-4 to -8%

Forecast Month	Nov	Dec	Jan	Feb	Mar	Apr	May	Jun	Jul	Aug	Sep	Oct				
Forecast Value	104	126	131	142	151	137	139	133	132	130						
Difference	4	6	6	7	1	2	4	6								60-day
% Difference	4%	5%	5%	5%	1%	1%	3%	5%						4%	3%	+1 to +7%

Forecast Month	Oct	Nov	Dec	Jan	Feb	Mar	Apr	May	Jun	Jul	Aug	Sep				
Forecast Value	112	136	145	152	165	152	150	140	140	140	135					
Difference	12	16	20	17	15	17	15	13								90-day
% Difference	12%	13%	16%	13%	10%	13%	11%	10%						12%	4%	+8 to +16%

84

In this example, the organization is still making single-point forecasts 30, 60 and 90-days in advance, but simple statistical analysis can predict, with about 95% confidence, that:

- The 90 day forecast will prove 8-16% optimistic

- The 60-day forecast will prove 1-7% optimistic, and…

- The 30-day forecast will prove 4-8% *pessimistic*

This assumes the forecasting process and demand seasonality remain unchanged, which is reasonable, since none show evidence of particular trending. Once improvement efforts begin, Excel's® "trendline" analysis of "% difference" will help gage success.

Hopefully, it is clear that having a min/max range of rational (quantitative) values for probable forecast-to-actual differences is very value-adding and worth the tracking effort.

Carefully consider the last bullet re supplier lead times. If a key item source needs 120 days to deliver to the organization, and the forecast only looks ahead 90 days, the Buyer/Planner is left to his/her imagination, and most organizations make ASWOs very uncomfortable. Surpluses *WILL* follow.

Tracking Flow

If you want to minimize inventories, you need the best possible flow of materials and products through the organization. Since what gets measured gets managed, good flow measurement is essential.

Obvious? Then why do so many organizations rely solely on Finance's "turnover" figures to gage flow?

To begin with, "inventory turnover" is a totally imaginary and artificial calculation… shipment-cost value (which includes labor, burden, and sometimes freight) divided by estimated average inventory (no labor, burden, etc, AND usually derived from last Inventory valuation – consumption since + new receipts since). At best this is "apples versus oranges", and offers little insight to actual, physical levels or activity. Its only value is for comparison to itself.

Getting away from "derived" inventory value requires (proper) cycle-counting, as described in a following chapter. Cycle counting ALSO avoids the trap of A-B-C distortions within derived values. For example, if I deal in gold coins, is my $500K of calculated inventory $450K of coin and $50K of packaging, or $50K of coin and $450K of packaging?

Getting away from misleading "shipment values" in the equation requires stripping away the non-material portions of the number and getting back to "shipment material value"… no labor, burden, freight, etc). Trying to do this precisely would be impractical, but most organizations know quite accurately what their usual labor/burden percentages are.

With all this "getting back to basics" in place, it is then relatively straight-forward to compose a 1-page "dashboard style" inventory flow-tracker that reports in approximate *days-of-supply*" units of measure, an Excel®-based example of which is shown on the next page.

There's a lot on this report, so step-by step review is in order.

First, notice along the left-most column of data labels, that the report is for a fully-integrated organization, and includes sales, finished goods, unimproved materials and components (WIP or Work-In-Process).

Along the top, you will note that each month has a column, with totals and averages to the right.

Then note there is a vertical divider line to highlight the "current dateline" and switch-over point between actual/historic values and estimated/future or business-plan values.

Input cells are indicated here by a dark background, but the actual spreadsheet makes extended use of fill color for ease of reading and comprehension. Note that there is an instruction at the end of this guidebook regarding how to request actual Excel® sample files for this and all other spreadsheets illustrated.

The key feature is the 4 rows, highlighted with block arrows, indicating approximate physical days of supply for the plan, actual months to date, and forward months.

Inventory Status Review

Unit: Sample Factory   Date: 31-May-09

Turnover (lines) (COGS/AVGINV)

	Plan 12-Pt	LE 12 Pt
	18.2	16.1

Physical & fiscal turns are '-

	Plan 12-Pt	LE 12 Pt
	18.2	16.1

Physical turns - above x 0.85

	Plan 12-Pt	LE 12 Pt
	9.1	8.1

	Jan	Feb	Mar	Apr	May	Jun	Jul	Aug	Sep	Oct	Nov	Dec	Total	Avg	Yr LE
				ACTUAL <		> LATEST									
Sales/Shipments (Invoice Value)															
Plan	500	600	700	800	800	1000	900	800	700	600	500	400	8400	700	8400
BOT F'cast	490	550	630	810	900	1000	1000	800	750	650	580	350	5050	721	5050
Actual	400	392	440	648	720								3430	686	3430
Cost-of-Goods-Sold (less Gross Material) [multiplier for COGS from Sales – 0.80]															
Plan	400	480	560	640	720	800	720	640	560	480	400	320	6720	560	6720
BOT F'cast	392	440	544	640	720	800	800	640	600	520	400	280	4040	577	4040
Actual	392	440	544	648	720								2744	549	2744
Material Sold (less Labor and Burden) [multiplier for Matls from COGS – 0.85]															
Plan	340	408	476	544	612	680	612	544	476	408	340	272	5376	448	5376
BOT F'cast	340	374	462	551	612	680	680	544	510	442	340	238	3232	462	3232
Actual	333	374	462	551	612								2195	439	2195
Finished Goods Inventory															
Plan total	264	317	370	422	475	528	475	422	370	317	264	211	4435	370	4435
BOT F'cast	300	350	400	450	500	600	500	475	420	400	350	300	3045	435	3045
Actual	300	350	400	450	500								2000	400	2000
Matched to Orders	40	45	50	60	80								275	55	
Unmatched < 30 Days	90	95	110	130	155								580	116	
Unmatched 30-60 Days	70	75	90	115	123								473	95	
Unmatched 60-90 Days	50	55	60	60	55								280	56	
Unmatched >90 Days	40	50	55	55	55								255	51	
Non-moving (>360 Days)	10	30	35	30	32								137	27	
FG Calendar Days of Supply (approx)															
Plan (min of current mo. + …)	18	18	19	19	19	21	19	19	18	18	18	18		19	
Actual or forecast (same)	22	21	20	20	20	23	19	20	20	21	23	31		22	
WIP Inventory															
Plan Total	264	317	370	422	475	528	475	422	370	317	264	211	4435	370	4435
BOT F'cast	300	350	400	450	500	600	500	475	420	400	350	300	3045	435	3045
Actual	300	350	400	450	500								2000	400	2000
In-Transit	10	12	10	15	17								64	13	
Non-moving (>360 Days)	15	16	19	22	25								97	19	
Other Fabricated IP	20	30	35	40	48								173	35	
Other Purchased IP	45	58	64	77	90								334	67	
WIP Calendar Days Supply															
Plan (min of current mo. + …)	21	22	22	22	22	25	22	22	22	22	21	21		22	
Actual or forecast (same)	24	24	24	23	23	28	23	25	25	27	28	26		25	
Total Inventory															
Plan Total	528	634	739	845	950	1056	950	845	739	634	528	422	8870	739	8870
BOT F'cast	600	700	800	900	1000	1200	1000	950	840	800	700	600	6090	333	6090
Actual	0	0	0	0	0	0	0	0	0	0	0	0	4000	333	4000

Seeing this overall layout is important, but difficult to read, so an enlarged section is depicted below.

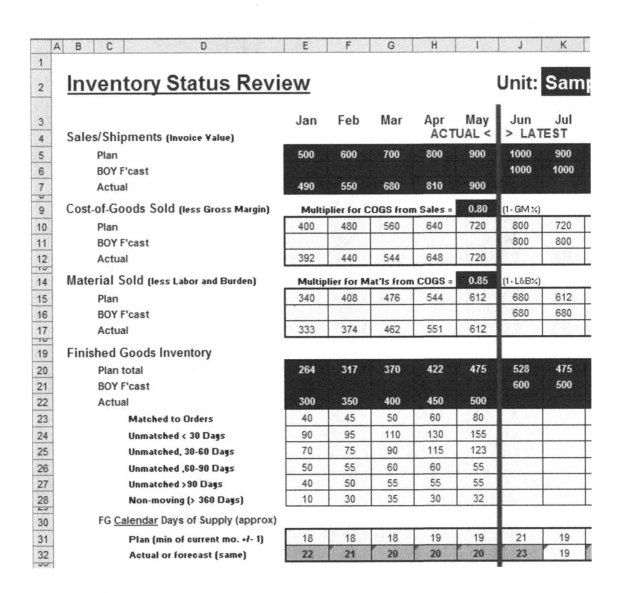

	Jan	Feb	Mar	Apr	May ACTUAL <	Jun > LATEST	Jul
Inventory Status Review						Unit: **Samp**	
Sales/Shipments (Invoice Value)							
Plan	500	600	700	800	900	1000	900
BOY F'cast						1000	1000
Actual	490	550	680	810	900		
Cost-of-Goods Sold (less Gross Margin)	Multiplier for COGS from Sales =				0.80	(1 - GM %)	
Plan	400	480	560	640	720	800	720
BOY F'cast						800	800
Actual	392	440	544	648	720		
Material Sold (less Labor and Burden)	Multiplier for Mat'ls from COGS =				0.85	(1 - L&B%)	
Plan	340	408	476	544	612	680	612
BOY F'cast						680	680
Actual	333	374	462	551	612		
Finished Goods Inventory							
Plan total	264	317	370	422	475	528	475
BOY F'cast						600	500
Actual	300	350	400	450	500		
Matched to Orders	40	45	50	60	80		
Unmatched < 30 Days	90	95	110	130	155		
Unmatched, 30-60 Days	70	75	90	115	123		
Unmatched ,60-90 Days	50	55	60	60	55		
Unmatched >90 Days	40	50	55	55	55		
Non-moving (> 360 Days)	10	30	35	30	32		
FG Calendar Days of Supply (approx)							
Plan (min of current mo. +/- 1)	18	18	18	19	19	21	19
Actual or forecast (same)	22	21	20	20	20	23	19

Rows 4-7 detail monthly shipment values (usual Finance top-line), in this case $(,000 omitted), but rows 5-6 switch from actual to forecast at columns I-J (May-June). These compare to row 5 (Business Plan), and on the actual spreadsheet, turn "cautionary pink" if below the plan value.

Rows 10-12 convert rows 5-7 to approximate "cost-of-goods-sold" using an organizationally-typical conversion factor of 0.8, as entered in cell I9 (Gross Margin is dropped).

90

Then, Rows 15-17 again translate to "Materials sold or shipped (dropping typical labor & burden) using the 0.85 multiplier input at cell I14.

Rows 20-22 itemize Finshed Goods inventory value, and 31-32 convert these to approximate days-of-supply as follows:

- Actual FGI book value ($500K in cell I22), which includes labor & burden but not margin, is compare to the estimated cost-of-goods-sold values in both cells I12 and J11 ($720K in the current month, and $800K forecast next month).

- (500 / 720) x 30 ~= 21 days of supply using current shipment cost values.

- (500 / 800) x 30 ~= 19.5 days of supply using next months forecasted shipments.

- Cell I32 reports ~ 20 days of supply (least value), allowing the Planner credit for watching and responding to apparent up/down trends.

Rows 23-28 are optional and customizable, but the choices shown here (FGI aged values) are highly instructive, and typically available

	A	B	C	D	E	F	G	H	I	J	K
29			FG Calendar Days of Supply (approx)								
30			Plan (min of current mo. +/- 1)		18	18	18	19	19	21	19
31			Actual or forecast (same)		22	21	20	20	20	23	19
33	WIP Inventory										
34		Plan Total			264	317	370	422	475	528	475
35		BOY F'cast								600	500
36		Actual			300	350	400	450	500		
37			In-Transit		10	12	10	15	17		
38			Non-moving (>360 Days)		15	16	19	22	25		
39			Other Fabricated IP		20	30	35	40	48		
40			Other Purchased IP		45	58	64	77	90		
42		WIP Calendar Days Supply									
43			Plan (min of current mo. +/- 1)		21	22	22	22	22	25	22
44			Actual or forecast (same)		24	24	24	23	23	28	23
46	Total Inventory										
47		Plan Total			528	634	739	845	950	1056	950
48		BOY F'cast			0	0	0	0	0	1200	1000
49		Actual			600	700	800	900	1000	0	0

Rows 43-44 show Materials days-of-supply estimates from a comparison of actual current materials value ($500K as in cell I36, ideally cycle-counted) to both current and close-forecast materials sold/shipped (cells I17 @ $612K and J16 @ $680K).

(500 / 612) x 30 ~= 24.5 days, and (500 / 680) x 30 ~= 22.1 days, so cell I44 reports ~23 days, again crediting Buyer/Planner trend maneuvers.

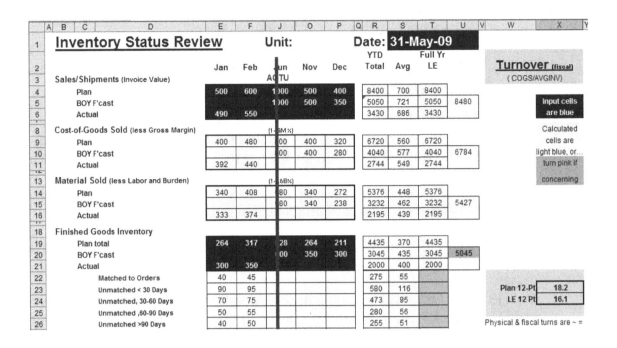

		Jan	Feb	Jun AC TU	Nov	Dec	YTD Total	Avg	Full Yr LE		Turnover (fiscal) (COGS/AVGINV)		
Inventory Status Review	Unit:					Date: **31-May-09**							
3	Sales/Shipments (Invoice Value)												
4	Plan	500	600	1)00	500	400	8400	700	8400			Input cells are blue	
5	BOY F'cast			1)00	500	350	5050	721	5050	8480			
6	Actual	490	550				3430	686	3430				
8	Cost-of-Goods Sold (less Gross Margin)			(1-|BMx)								Calculated cells are	
9	Plan	400	480	)00	400	320	6720	560	6720			light blue, or...	
10	BOY F'cast			)00	400	280	4040	577	4040	6784		turn pink if	
11	Actual	392	440				2744	549	2744			concerning	
13	Material Sold (less Labor and Burden)			(1-|&Bx)									
14	Plan	340	408	80	340	272	5376	448	5376				
15	BOY F'cast			80	340	238	3232	462	3232	5427			
16	Actual	333	374				2195	439	2195				
18	Finished Goods Inventory												
19	Plan total	264	317	28	264	211	4435	370	4435				
20	BOY F'cast			)00	350	300	3045	435	3045	5045			
21	Actual	300	350				2000	400	2000				
22	Matched to Orders	40	45				275	55					
23	Unmatched < 30 Days	90	95				580	116			Plan 12-Pt	18.2	
24	Unmatched, 30-60 Days	70	75				473	95			LE 12 Pt	16.1	
25	Unmatched ,60-90 Days	50	55	.			280	56					
26	Unmatched >90 Days	40	50				255	51			Physical & fiscal turns are ~ =		

Finally, all totals, averages and latest full-year estimates are displayed to the right, and converted to turnover metrics (both Plan and LE versions).

In the case of Finished goods, Fiscal turns and physical turns are about the same, while the materials or WIP turns (cells X37 & X38) are physical estimates, and only about 85% of what the Finance group might report, inflating material shipment value by the labor and burden factor.

The key point is that by keeping the inventory status report simple (1-page), and offering readers a choice between traditional $ or days-of-supply, plus a choice between fiscal or physical turns, people can better visualize and "get their mental arms around" the information.

For my way of thinking, "20" days of supply is easier to comprehend and respond to than "$500K" of inventory, and with a little thought, takes no more time and effort to report than current typical reporting.

S&OP teams, Inventory/Materials Czars and Value-Stream Managers all like such systems very much once they are tried.

A-B-C Materials Classification and Management Strategies

A lot of organizations *THINK* they have and use an A-B-C materials classification system.

Closer examination too often discloses that only some small portion of materials bear an A-B-C tag, that nobody knows when it was assigned, how, by whom, when, who uses it, whether it has ever been updated, or even its purpose.

Typically it is the remnants of some long-retired IE's since-forgotten attempt to classify materials for relative frequency of use, and packaging/storing in a way that minimized material handling time and costs.

Such logistics-focused systems frequently break down because the target is constantly changing, physical spaces are not as easy to constantly change, benefits are difficult to quantify, and nobody is assigned the system-maintenance responsibility.

What I propose (and have used to very good effect) is an entirely different kind of A-B-C classification system... one that is relative-usage-*VALUE* based, that is targeted on improving Buyer/Planner productivity & effectiveness, that makes achieving targeted turnover *MUCH* easier, and which lays the foundation for proper cycle-counting and on-hand quantity data accuracy (fewer ASWO's and surpluses).

Almost inevitably, the Buyer/Planner team seems too small, overworked, chaotic, and in spite of sincere hard work, ASWO's and surpluses are a daily reality. These people are further handicapped by inappropriate data base settings and questionable on-hand quantity accuracy.

A big part of the problem is that there is not, and never will be enough time for the Buyer/Planner team to micro-manage every item, and the proposed solution is an A-B-C classification system that harnesses Pareto's Law to rebalance time and work:

A-B-C Classification
Pareto's Law at work: working smarter vrs. harder

Class	% Items	Model % Usage $	% Usage $ (Typ.)	
A	10	70	80	← Then, focus time & effort here
B	20	20	15	↑
C	70	10	5	→ Find "auto-pilot" mgmt methods
	100	100	100	

		A	B	C
Material Class =				
% of Value =		80	15	5
Est. WIP Turns (fiscal) =	36.7	51.4	17.1	6.4
Days of Supply (physical) =	11.6	7.0	21.0	56.0
Weeks of Supply (physical) =	1.7	1.00	3.00	8.00
		↑	↑	↑
		10% Items	20%	70%

Simply stated, Pareto's Law says that in a normally varied and distributed collection of items or events, only about 10% will account for the majority (70% +/-) of some valued characteristic, that another 20% will account for an additional (20% +/-) portion of the valued characteristic, and that the majority of items or events (70% +/-) will account for the smallest portion (30% +/-) of the valued characteristic.

This seems mysterious, but it works and is quite trustworthy. What % of your monthly household expenses are due to largest 10% of the bills?

Happily, when focused on item relative usage-*VALUE*, this phenomenon can be used to manage more materials, better, with less time, effort and confusion!

In the above example, a quick query on the site MRP disclosed that 10% of the items represented 80% of their USAGE VALUE, another 20% accounted for an incremental 15% of USAGE VALUE, and that the remaining 70% only accounted for 5% of the usage value. By itself, this knowledge means they could probably devote most of their time to micro-managing just the "top 10%" if items, and be close-controlling 80% of the system value!

Taking the calculation even further (method details will follow), it can be shown that they could reliably look forward to over 37 turns by simply keeping A-B-C weeks-of supply under 1-3-8 (1.7 usage-value-weighted average). Think about this! 37+ turns, in spite of holding ~8 wk-of-supply on 70% of items!

So how does such a system operate?

A-B-C Classification

Process & Toolkit for Maximum Controlled Turnover

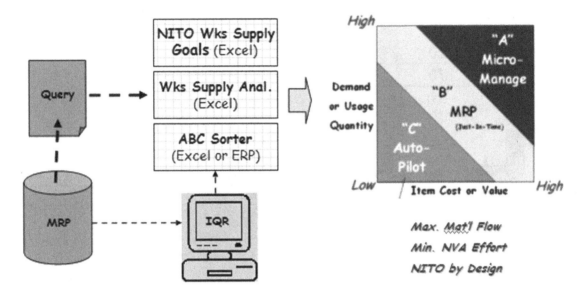

It starts by writing a (re-usable) query on the materials management system (MRP or DRP). This automatically takes a "snapshot" of key data elements whenever the system is fully updated (usually daily or weekly), and makes it possible to use and play with the data without bogging down or interfering with that system.

This data can then be inserted to various Excel® spreadsheets for analysis. Note the plural on "spreadsheets"! As long as the query is going to be run, collecting ALL interesting data enables MULTIPLE uses of the same snapshot without the time and effort to impose several queries on the system.

Spreadsheets that are recommended and will be illustrated include:

- A relative-usage-value-based A-B-C item classification tool/sorter

- An item on-hand-weeks-of-supply calculator that ALSO helps identify suspicious or troublesome data points or settings, and...

- A goal-setting calculator that converts annual turnover goals to A-B-C weeks-of-supply targets (and vice-versa).

Armed with the information from these 3 spreadsheets, Buyer/Planners can quickly correct questionable system settings, select an appropriate management method or strategy for each item, and focus their attention on items that really matter financially.

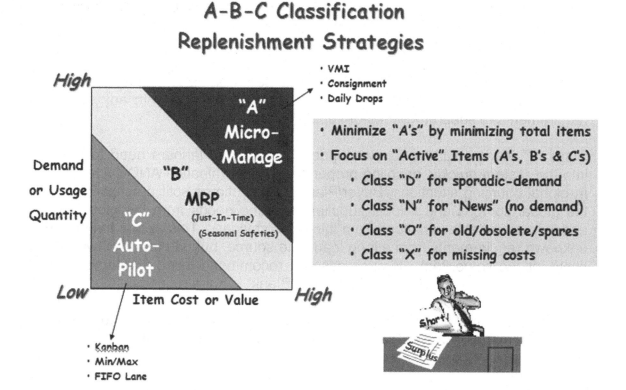

A-B-C Classification Replenishment Strategies

Here is a standard "strategy guide" that has worked well in a number of organizations. In this case, A-B-C classifications are only assigned to items that get *some* amount of regular use. This keeps the collection smaller (important to B/P time management AND cycle-counting (discussed in the next chapter). A class D is assigned to sporadic demand items (very infrequent call), a class N for new items (no forecast or history yet), a class O for known-to-be obsolete or replacement-business-only items (special rules), and a class X for items with missing unit costs (A-B-C, but usage value cannot be determined).

"A" items get micro-managed... set for daily deliveries, or become the subject of Vendor-Managed-Inventory methods (such as consignment).

"B" items are managed via "just-in-time" methods, possibly including supplier-supervised system levels, and typically rely on (seasonally adjusted) safety stocks (again, ideally held by suppliers).

"C" items are set up for minimum-attention, "quasi-autopilot" management methods such as kanbans, min/max systems, or FIFO Lanes (fixed stock area that the supplier or back-shop watches, never allows to go empty, and always tries to keep full).

Class "D" items are procured-to-order only, possibly with a safety stock equal to the largest, single, historic need (depending on item cost and lead time).

Class "N's" are treated as "A's" until demand data allows A-B-C classification. The "N" helps the Buyer/Planner watch for this transition and possibly do the reclassification without having to redo the entire item set.

Class "O" designation helps Buyer/Planners clear non-value-adding material, and to maintain replacement component sock levels in keeping with any contractual requirements or separate weeks-of-supply targets.

Class "X" items could be "show-stoppers", so Buyer/Planners need to get costs into the system *quickly*, enabling proper A-B-C determination AND cycle-counting. In global sourcing situations, Buyer/Planners sometimes split purchases between the global sources and a local supplier (baseline needs from the global source, and less predictable variations from the local provider). This avoids trying to meet unknown requirements from a long lead-time source, but can result in two or more costs for the same item. In such cases, I recommend using a volume-weighted average unit cost for A-B-C classification. "Blank" is not a responsible option.

Local strategy maps can vary. The point is that having such a (documented) approach improves Buyer/Planner choices, consistency, effectiveness and time-management.

If you are not convinced, read the beginning of this chapter again, and ask yourself what viable alternatives you have. If you ARE convinced, putting such a system into place requires collecting the needed data and creating or selecting the Excel® processing spreadsheets.

Record Layout for ABC & Cycle-Counting Classification Analysis

- The IWO data input file should be composed from your MRP (WIP) or DRP (FG) database and contain only ASCII characters in a delimited format as defined below, and easily parsed/read into Excel
- Include all parts or SKUs with a balance on hand greater than zero, and/or parts with future requirements, and/or parts with use in the last 12 months, and/or parts with an open or planned purchase order or manufacturing work order.
- Unit Cost (Field 11) should include the decimal point (or equivalent), and up to a maximum of four places. Use of international, regional standards for currency is acceptable... whatever is in your MRP database.

Field No.	Field Name	Type	Comments
1	PART NUMBER	a/n	
2	DESCRIPTION	a/n	
3	PURCHASE / MANUFACTURE	a	P or M (use capital letters)
4	PLANNER CODE	a/n	
5	BUYER CODE	- -	

This guidebook's Appendix contains the MRP or DRP data snapshot layout details needed to support the analysis spreadsheets depicted. If passed to your I/T team, they should be able to create the needed query in a matter of hours and schedule it to run every time the system is fully updated (latest completed shipments, orders, deliveries, schedules, etc). Their motivation for helping is the to-follow, significant reduction in special-query requests AND variable query time demands on the system.

This leaves the challenge of creating or "borrowing" the Excel® analysis spreadsheets.

If you like the following examples, the Appendix provides instruction on how to receive complimentary copies… if only as starting points for developing your own.

Excel®-based A-B-C Assessment Tool

- Data from MRO or IQR Query

- Max 30-day demand value from 12 months past and 3 months forward

- Records sorted per decreasing max 30-day demand $

- Cum % items and cum % demand value are tallied

- A, B, C types calculated per 10-20-70 % of items

- (Type Z for demand value = 0… no demand or no std cost)

- Current A-B-C setting highlighted if different than calculation

Here's the overall layout of an A-B-C item classification spreadsheet. Don't worry if it is blurred… Close-up images follow.

The point of this image is overall layout and usage steps.

The item data from the MRP or DRP snapshot is read into an Excel® table, then copies into the first (21) columns of the spreadsheet. Only about half gets used by THIS spreadsheet, but the rest will get used by others, and getting it all at once avoids having to create and repeatedly run several different queries.

Once "loaded", the spreadsheet calculates maximum 30-day demand value within a 15-month window (past 12 months + planned 3 months future). This eliminates the confusion-creating effects of seasonality and short-term trends.

All items are then user-sorted in decreasing order of 30-day demand value, and the spreadsheet calculates/reports the cumulative % of items listed and the corresponding, cumulative % of total 30-day demand values. THIS enables current A-B-C groupings and value-distribution reviews, by "inspection". The first 10% of items are designated as "A's", and their % demand value can be found by looking for the A/B split in the list. The next 20% are designated as "B's" and the remaining 70% as "C's". Items with missing unit costs or no 15-month demand are tagged as "Z's" or "X's".

Finally, the spreadsheet reports any existing A-B-C classification from the MRP or DRP system, and highlights any differences between existing and "fresh" classifications.

For a 30,000 item system, full processing and analysis can be done in under 15 minutes, and follow-up time gets shorter as the data base integrity is improved.

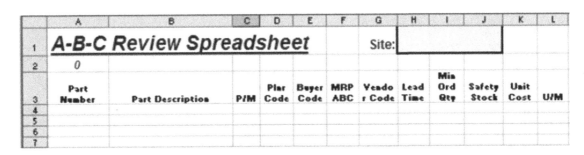

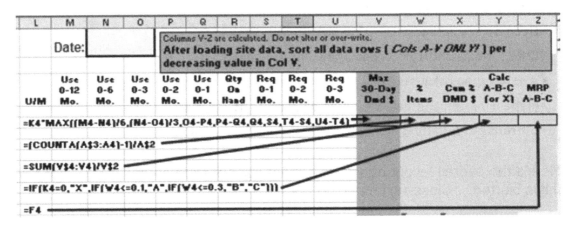

Here's some images large enough to read, and including the row/column headers so that cell formulae can be understood.

Remember, instructions for obtaining an actual sample of this (and other featured) spreadsheets are in the Appendix. It would have been handy to include a CD or USB memory with the guidebook, but book publishers charge big money for such special features, and having to request the files puts us in touch with each other.

Great! You now have a relatively fast and reliable way to create and maintain a coherent usage-value-based A-B-C item classification system, but how will Buyer/Planners know what weeks of supply targets to attach to each class?

How will they know what weeks-of-supply class targets will combine to achieve the overall, business plan turnover goal?

Weeks-of-Supply Goal Calculator
RM&C Turns Goal known, Weeks-of-Supply Targets Needed

© ILSD llc, 2009

INPUTS
OUTPUTS

Site Labor & Burden % = [15] (L&B as % of Cost of Completed Production or Cost of Goods Sold)

		A	B	C	
Material Class =		A	B	C	
% of PN's =		10	20	70	
% of Value =		70	20	10	(Typically 70-20-10)
RM&C Turns Goal (fiscal) =	26.0	45.1	14.4	6.3	
Days of Supply (physical) =	16.3	8.0	25.0	57.1	
Weeks of Supply (physical) =	2.3	1.14	3.57	8.15	

Weeks-of-Supply targets known, Turns Impact Needed

		A	B	C	
Material Class =		A	B	C	
% of PN's =		10	20	70	
% of Value =		70	20	10	(Typically 70-20-10)
Est. RM&C Turns (fiscal) =	26.3	51.4	12.9	6.4	
Days of Supply (physical) =	16.1	7.0	28.0	56.0	
Weeks of Supply (physical) =	2.3	1.00	4.00	8.00	

Here's a spreadsheet for answering those questions, and which is also available on request.

In the top portion, the analyst enters the overall RM&C (Raw Material and Components) turnover goal (26.0),the site specific "% of Value" for each class (from the A-B-C spreadsheet, in this case 70, 20 & 10%), and the overall Site Labor & Burden % (here 15%).

The spreadsheet then reports A-B-C weeks of supply maximums that will combine to reach the specified turns (1.14, 3.57, 8.15).

Since these goals may be awkward to remember or put into daily use, the bottom portion of the spreadsheet allows entering simpler goals (1, 4, 8), and seeing the resulting overall turns implies (26.3).

Clearly, this tool was created for a manufacturing environment. The "% Labor and Burden %" is included to reduce FISCAL/Financial turnover (which includes L&B)

to *PHYSICAL* quantities (materials weeks of supply).   In a finished products distribution situation, set the L&B% to zero (product gets shipped at received value), or to the typical % markup (shipped value versus acquisition value).

This tool is a little deceptive. There is quite a bit of hidden "black magic" in the calculations, and a sadistic university Professor would "leave derivation of the details to the interested student" (because it is a question in the final examination). Fortunately, I'm an Engineer rather than a Professor:

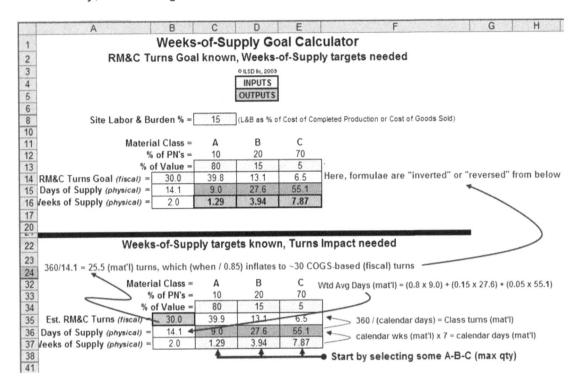

Now that item A-B-C classes and wks-of-supply goals are known, the last question is how many wks-of-supply are currently on hand?

Here, a 3rd spreadsheet comes into play (and a sample can be requested).

This spreadsheet takes the same "same snapshot" used for A-B-C classification, and calculates on-hand weeks-of-supply for each item... but with a "twist" in the logic used...

- The spreadsheet first determines the maximum 30-day demand quantity over the last 12 months and the forthcoming 3. Again this eliminates seasonality and stabilizes calculations if done frequently.
- The following calculation is "Minimum Wks of Supply", using current on-hand quantity, maximum 30-day demand, and the notion that 30 days is ~ 4.3 weeks.

- This approach acknowledges the idea that stock levels can seldom accurately track increasing or decreasing demand, and that enlightened management includes keeping changes gradual. Certainly "B & C" items need management systems that embrace maximum demand.

Note also that attaching "filter buttons" to each column enables this spreadsheet to function as a "questionable or troublesome" systems data detector!

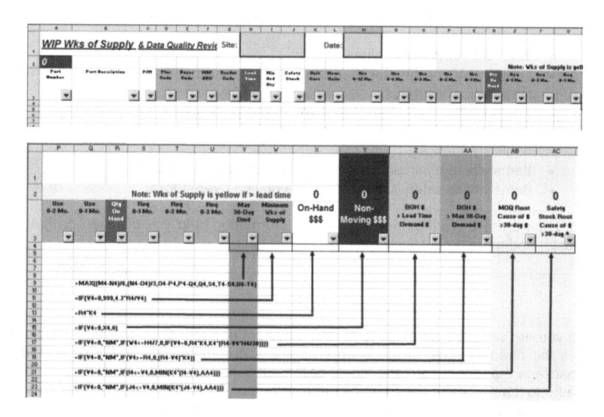

For example:

- Does column B contain any missing/blank part descriptions which complicate Buyer/Planner deliberations?
- In column C, which items are purchased, and which are created internally?
- Do columns D & E show any items without Buyer/Planner Code assignments? If so, who's in charge?
- Does column G show purchased parts without vendor codes? Oops!
- Column H: What are the longest lead times? Any set to "zero days lead time" (guaranteed future ASWO)?
- Any items in columns I or J with minimum order quantities or safety stocks set bigger than A-B-C wks-of-supply target maximums?
- Any missing unit costs in column K (blocks A-B-C classing and cycle-count planning)

- Column L: What items have which units of quantity measurement (for Inventory and cycle-counting tools/methods planning)?
- Columns M-Q: Any negative usages? They DO occur. Trust me.
- Columns S-U: Any negative forward requirements or forecasts? They ALSO occur!
- Column R: Any apparently negative on-hand quantities (already discussed)?

The spreadsheet then goes on (columns X-AC) to calculate and report:

- Total On-Hand Item Values
- On-Hand Value greater than the value of maximum needs during the lead time (logically surplus)
- Any non-moving on-hand value (no demand, obsolete, excess)?
- On-Hand Value greater than the value of maximum needs during 30 days (logically surplus)
- Items where minimum order quantity is driving over-stocks
- Items where safety stock setting are driving over-stocks.

If you were a Buyer/Planner, would this information help you do your job better, faster and more productively?

If you are seriously attracted to having a system and management capability such as the one described in this chapter, there is a commercially available software package that does it all (and more) called IQR® (Inventory Quality Ratio). I often refer to it as "A-B-C on Steroids", and have found it easier to use and less prone to mistakes than spreadsheets. It is also much easier to teach to Buyer/Planners, and allows "free-form" queries & reports from the system database snapshot.

This IQR® package is described in the Appendix, and can be more fully investigated at www.IQR.com.

Remember: If you want to do proper cycle-counting (next chapter), you **must** start with a demand-value-based A-B-C classification systems, toolkit and process.

Cycle-Counting (versus Stock-Taking)

I can *not* think of anything *more* likely to cause both ASWO's and surpluses than an incorrect MRP/DRP management-system quantity or balance-on-hand number!

So why on earth do so many organizations depend on a simple, annual full stock-taking to "recalibrate" these numbers... especially since a 15% error rate is common during the counting, and accuracy can degrade to just 40-50% in just weeks or months?

It is no secret that there is an alternate process called "Cycle-Counting" which, if implemented properly, is capable of 99% accuracy, but few make the commitment:

- They think they do it, but in reality only occasionally send somebody to the floor or warehouse to double-check on a suspicious system-reported quantity, or following an ASWO. This is not systematic cycle-counting!

- They have investigated it, but come to the (erroneous) conclusion that it takes too many people and costs too much.

- The people needed to do it don't produce any product (NVA but necessary), are classified as "indirect labor", and are usually the first to be let go when economic times get tight. Before exploring why all these excuses are nonsense, it is important to fully understand why quantity-on-hand accuracy is such an absolute need:

Cycle-Counting

Systematic, continuous counts vrs. major, all-items stock-taking

U.S. Sarbanes-Oxley Act
- CEO & CFO accountable for $ accuracy
- All reasonable efforts & methods required
- Up to $1M and 10-yr penalties

IRA Key to Operations
- Inventory Record Accuracy (IRA)
- Minimize Item Shortages
- Minimize Item Over-stocks
- Trust in MRP data
- Enables A-B-C strategies

Traditional stock-taking IRA ~= 50-70%

Cycle-Counting IRA ~= 95%+ with QCPC & RRCA

Most Auditors will allow C-count vrs. Physical Stock-taking once it demonstrates 95%+ IRA

From an operational standpoint, the benefits are entirely predictable, AND, once IRA (Inventory Record Accuracy) exceeds 95%, most Accounting Firm Auditors will allow you to discontinue annual or bi-annual stock-taking… a HUGE savings.

In addition, cycle-counting ensures regular checks on the quantity, so stock-taking's "accuracy erosion over time" goes away.

If you operate in the USA, the most compelling motivator is that your CEO and CFO can do jail time if they know their inventory numbers are grossly incorrect yet do nothing to improve them!

Cycle Counting

Process & Toolkit for Maximum "IRA" & Controlled Turnover

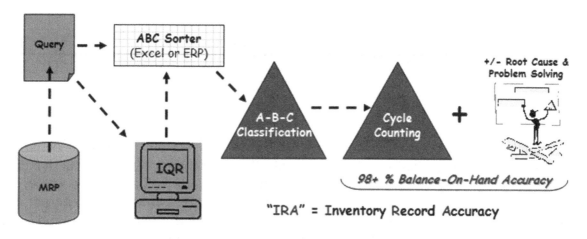

"IRA" = Inventory Record Accuracy

The key elements of a proper cycle-counting process are depicted above:

- MRP/DRP query data is used to create, and periodically update, a usage-value-based A-B-C classification system (see the previous chapter). If you do not have such an A-B-C system, you cannot implement proper cycle-counting.

- The A-B-C classifications are used to develop a cycle-counting plan, and each item is checked for actual vs. system quantity-on-hand at least 1-4 times per year, depending on its classification

- Any discrepancies are "autopsied" for root-cause(s), and lasting corrective action is implemented… not for just the item, but for the general root causes. In this way, the number of discrepancies become steadily fewer, and the CC-Team can be down-sized as appropriate.

"IRA" = Inventory Record Accuracy

Count (100) item-locations

If 90 are within tolerance, IRA = 90%

"Hit" or "Miss"... No "close-enough's"

Class	% Items	Normal Tolerance	Sample Items	Ideal Hits	Design Hits
A	10	+/- 0%	50	50	50
B	20	+/- 2%	100	100	98
C	70	+/- 5%	350	350	332
	100		500	500	480
	IRA = Hits / Samples =			100%	96%

A variable set of counting tolerances builds on the (Pareto) nature of a usage-value-based A-B-C system. A's are only 10% of the population, but typically represent 70-80% of the usage-value (and inventory valuation accuracy), so proper cycle counting allows no actual/system quantity tolerance. By contrast, C's are 70% of the population, but only represent 5-10% of the usage value, and up to 5% tolerance may be acceptable, depending on overall accuracy goals.

The other practical consideration involves units of measure and method of quantity measurement. For example, perhaps "titanium ball bearings" are A-items, BUT counting them individually is impractical, so weighing of containers is used as a proxy for item counting, and accuracy must be within the error of the scale.

Again, a key benefit of proper cycle-counting is achieving 95%+ IRA, and persuading the Auditors that annual stock-taking is no longer needed, so tolerances *MUST* be kept as tight as is practical.

By the same token, counting cannot be random, and auditors will look for proper team sizing, count frequency and discrepancy resolution.

Cycle-Counting
Successful Planning Parameters

Class	% Items	% Usage Value	# Site Items	Minimum Counts/Yr.	Benchmark Hrs./Count	Req'd Hrs/Yr
A	10	70	2000	4	0.5	4000
B	20	20	4000	2	0.5	4000
C	70	10	14000	1	0.5	7000
			20000	30000		15000

Req'd Hrs/Week = 300

Team Size ~= | 9 |

15-20 Min. / Item (counting)
15-20 Min. Admin & QCPC/RRCA
30-40 Min. Avg. per Item counted

Note: Successful processes (IRA improves) budget 50% of team time for problem solving

Seem costly? What's the avoidable current $ cost of:
- Annual physical inventory events?
- Line shutdowns (and pre-builds) for inventory events?
- Safety stocks and stock levels driven by poor IRA?
- Item stock-outs and shortages ("ASWO's")?
- Materials expediting and emergency freight charges?

Think "COPIRA": (total) Cost Of Poor Inventory Record Accuracy!

Item count frequency depends on A-B-C classification. A's should all be counted at least 4 times per year, B's at least twice, and C's at least once. If IRA is slow to improve, frequency (and team size) may need to grow until the targeted improvement rate is observed.

A reasonable time allowance for counting an item is about half an hour... 15 minutes +/- for actual counting (assuming proper tools and location system), plus an average of 15 minutes +/- for discrepancy resolution. Obviously a "match" between actual and system quantities requires no resolution time, while a "miss" can require more than 15 minutes to resolve. It will also be observed that the number of discrepancies will shrink as root-cause resolutions take effect.

In the example above, a collection of 20,000 stocked items require 30,000 counts per year, 15,000 Team-member hours, or about 9 Team members. This appears to be 50% more counts than are required in a traditional, annual stock-taking, but remember, IRA can be expected to go from 50% to 95%+ in return, operational shutdowns go away, no shutdown advance production is needed, safety stocks and stocking levels can come down, ASWO's can all but disappear, and expediting/emergency-freight can follow suit.

Think "**COPIRA**" (total **C**ost **O**f **P**oor **I**nventory **R**ecord **A**ccuracy), and you should quickly conclude that you cannot afford to NOT implement proper cycle counting. In fact, it is a good idea to calculate a baseline COPIRA value, and track it as a means of measuring CC program progress... just like **COPQ** (total **C**ost **O**f **P**oor **Q**uality).

It is also useful and advisable to learn from those who have both succeeded and failed at cycle-counting implementation.

Cycle-Counting
2002 APICS Member Study

Best Practices...

· Counting every day

· Std Work for choices & methods

· Bar coding to minimize data errors

· Count by location (no "netting")

· Use 3rd-Party staff (no bias or headcount)

· Include "IRA Charts" in mgmt. metrics

· Rational plan & system

and Worst Practices...

· "Flexible" class tolerances

· Resetting system qty. without problem-solving

· Variable data collection schedules

· Random item selections (vrs. class-based)

· "Secret" results

· "Spare Time" or "Expediting" activity

· Untrained team and supervision

These seems straightforward and logical, but some bear emphasis.

CC is a full-time, *EVERY*day activity. Trying to do it 1-2 days a week leads to unreliable team composition, delayed resolution actions, and second-level priorities.

A standard, documented (automated?) method is needed to specify the items to be counted, and to ensure minimum annual frequencies. The fact that somebody questions a system quantity is good, but NOT a CC item-selection criteria, and the CC Team should not redirect its planned investigations or time.

If "indirect headcount" is a concern, consider using an outside service agency. Be certain, however, that they use the same, trained people and Supervisor every visit. More difficult, but as important, is ensuring their contract is not "expendable" during economic dips…. Interrupting requires starting over!

Failing to uncover and correct discrepancy root causes means IRA will NEVER improve.

Keeping IRA% "secret" invites loss of the CC program budget. Make certain IRA is getting regularly better, and make sure the whole organization knows it.

As with any major change, resistance is predictable, and graduated introduction may need to be employed…

Cycle-Counting
Tips for successful introduction

Success Factors:
- Trained & committed Exec Sponsor
- Identified current co$t of poor-IRA%
- IRA-accountable Manager
- Trained, ongoing team members
- Firm 95% + IRA Goal
- Rewards for improvement
- Adequate staffing
- Public results
- Urgent actions toward site-wide deployment and robust A-B-C system linkage, but…
- *Start with a "control group" of items*

Control Group?
- As large as possible, but…
- 33 items minimum
- 5% A's, 10% B's, 35% C's, 50% known problems
- Keep the same items until each has a "clean" record for 10 counts, then replace with a like item (A, B, C, problem)
- Focus on RRCA and Problem-Solving processes before attempting 100% deployment.

These ideas also come from APICS case studies and conference presentations.

The cycle-counting Team and Team Leader need to be trained, made accountable, rewarded for quantified success, equipped with standard procedures, and kept intact. Trying to do the job part-time or with random team membership makes training, focus and motivation practically impossible.

If Management balks at fielding a correctly-sized team at first (see previous team-sizing calculations), the process's introduction may need to be scaled to the number of people available, and focused on a "control group" of inventory items.

This control group of inventory to be targeted should contain at least 33 items (for statistical validity), and include at least 50% "known bad actors"… items the Buyer/Planners know, from experience, are seldom in calibration. This is because the emphasis needs to be on Relentless Root Cause Analysis (RRCA) and (permanent) problem-solving or mistake-proofing.

It is also recommended that the Team develop a "standard operating procedure" for their work, so that they can become as efficient as possible, and more easily train new members. The one that follows is a composite of SOPs used at For Motor Co. and UTC, and seems to work well:

Cycle-Counting
Counting is just the 1st step, and only half the job!

DIVE Step	8D Step	Problem Solving Step	Purpose
D	D1	Select the team	Establish a work group with knowledge and/or experience to solve a particular problem and implement corrective actions.
	D2	Describe the problem	Describe a product or process problem to answer in measurable terms Which, What, Where, When, How, and Why.
I	D3	Implement and verify interim containment measures	Reduce or eliminate a customer's risk of experiencing the problem by quickly implementing actions or safeguards.
V	D4	Define and verify root causes	Identify the root cause of the problem
	D5	Identify and verify corrective actions	Identify and verify a corrective action to permanently reduce or eliminate the problem
E	D6	Implement permanent corrective actions	Define and execute an implementation plan for the corrective action.
	D7	Prevent recurrence	Verify the stability of the corrective action over time.
	D8	Recognize the team and report out	Recognize the efforts and success of the 8D team. Report results to customers.

112

And to "prime the pump", here is a useful (but only partial) list of IRA/count "miss" root causes that will almost inevitably surface and beg operational changes and/or mistake-proofing.

Discussing these and brain-storming others, in advance of resolution process efforts, can save time and sharpen team member's sensitivities...

Cycle-Counting
IRA "Misses": Common Root Cause Findings?

- ❏ No quantity & quality-verification at Receiving
- ❏ BOM quantity errors
- ❏ Incorrect count on last inventory or check
- ❏ Mixed items in locations
- ❏ Missing or illegible item labels
- ❏ Item substitutions & confused supercessions
- ❏ Inadequate location system
- ❏ No location-system training or discipline
- ❏ Confused measurement units
- ❏ Undeclared schedule changes & "pull aheads"
- ❏ Unauthorized and undeclared work-order batching
- ❏ Delivering more/less than needed by Ops (pallet qty's)
- ❏ Delayed or missing system transactions...
    - ➢ Receiving
    - ➢ Defects or damage
    - ➢ Substitutions
    - ➢ Moves
    - ➢ Re-stocks / returns to stores from Ops
    - ➢ Partial-builds (missing a component)
    - ➢ QC holds
    - ➢ Location system changes or re-assignments

Finally, and as always, "the devil lies in the detail", so what follows are some of these details, as offered to me by a successful, cycle-counting process implementer and Program Manager...

C-C's objective is to validate planning & control system (MRP or WMS) quantities with actual, so what gets counted should match what the "system" is describing:

- If the "system" is reporting a single quantity, in a single location, or multiple locations with separate quantities, each should be cycle-counted and scored individually. Finance doesn't care, but Operations benefits from having predictable quantities in specific locations... (but see the note below re points of use and item count timing).

- If the "system" is reporting a combined quantity for several locations (e.g. Receiving, Paint, Central Stores, and 3 different Supermarkets), all locations need to be counted and the total scored against the system value. This WILL take more counting time, so re-think your time & staff budget!

- Where the system only tracks "overall" quantity, the team will need a reliable guide to all points of storage and use for all parts (suggested content for any PFEP: Plan-For-Every-Part)

- The question of timing is important. The "system" is reporting quantities at a specific point in time... usually in the range of midnight to 5AM, when the system is updated for all activity the previous day. This means that if additional activity has occurred between system-update and the cycle-count (e.g. production, location changes, etc), the comparison of system and actual won't be immediately obvious... which is a key part of the "rationalization process" in the time-budgeting model. Some teams ease the effort by doing counts on the 3rd shift, and comparing them to a "live" (updated) system terminal.

- Another typical question involves how to treat items on a production line and in process of being consumed. If the system count was reduced when the materials were delivered, they do not get counted. If the system count is only reduced when final production is reported (back-flush), the team needs to get an accurate idea of the "moving target", and possibly even make count-tolerance adjustments for such materials.

- Buyer/Planners will immediately try to get the C-C Team to do (expediting) counts. *Politely decline.* "Cooperating" will upset item selection, control group, and staffing plans... become a daily nuisance task, and take much longer than expected (remember, such requests are based on problems!)

Summary

Although it typically involves a full-time, indirect headcount team and 50% more item counts than traditional, annual stock-taking, proper Cycle-Counting can take on-hand quantity accuracy from 50-60% to 95%+, eliminate the need and associated costs of annual stock-taking, drastically reduce both ASWO's and surpluses… and help keep your Senior Execs out of prison. It may appear somewhat costly at first, but calculating and tracking "COPIRA" (total Cost Of Poor Inventory Record Accuracy) rapidly points to an outstanding return on investment.

Thoughts on Using 3PL Services

The term "3PL" was defined earlier, but in case you missed it...

A "3PL" is a 3rd-Party Logistics provider... a "supplier" who provides you with one or more material and inventory logistics services so that you are free to focus on your "core competencies" (fabrication, assembly, etc.)

Some years ago, I was fortunate to be part of a 3PL services benchmarking team, and was invited to visit and see how a major, would-be 3PL provider was fully supporting a well-known auto and truck manufacturer's assembly operations.

What I witnessed was literally amazing:

- The 3PL had occupied a well-designed warehouse immediately adjacent to the auto assembly plant, and staffed it with a dedicated team of highly-trained, career professionals.

- Each Wednesday, the assembly plant would provide a firm, vehicle-by-vehicle assembly schedule and sequence for the week to come, and a tentative sequence for the following week.

- The 3PL participated in the assembly plant's S&OP meetings, and took responsibility for component planning and just-in-time procurement from hundreds of suppliers, both independent and auto-maker owned.

- The 3PL handled all receiving and staging, and supplemented the assembler's inbound QC team with its own, and provided professional packaging/damage-control expertise.

- The assembly plant operated 2 shifts, so the 3PL converted the vehicle assembly schedule and sequence to vehicle-specific component kit requirements (including customer options) for each assembly station, then delivered 4-hrs of properly sequenced kits 4 times per day.

- The 3PL maintained emergency safety component safety stocks to protect against damage situations

- Key, contracted performance metrics included:

    o Overall, non-material costs per vehicle

    o On-hand materials days-of-supply and annual turnover

    o Inventory Record Accuracy (99%+ goal)

    o Availability of kits where and when needed

    o Parts-per-million damage and mitigation costs

- The joint operation was consistently achieving 99%+ on all performance goals

Obviously, the 3PL in question was one of the best, possibly World-Class, and the folks in the Assembly Plant's operations group understood the process and it's needs very well… including planning, scheduling, purchasing, quality, and communications (both electronic and human).

So just what services can a 3PL potentially provide?

- Procurement planning & ordering (you usually still pay the suppliers)

- Freight Management

- Receiving & staging

- Material "de-trashing" and re-packaging

- Materials labeling

- Inbound quality control

- Warehousing

- Component kitting

- Delivery cart design, construction & maintenance

- Physical delivery to specified points of use

- Materials location control and on-hand quantity records accuracy

- Expert advice & counsel re the logistics needs of target process design, new product introductions, volume changes and supply-chain/sourcing improvement efforts.

- Full, finished-goods packing and outbound freight management services.

Essentially, a professional, experienced 3PL provider can do whatever you want or need, BUT the more you want, the greater the effort required by everyone to achieve it. *NONE* of the services are simply "off-the-shelf", "plug-and-play" or "free".

If you are attracted to the 3PL options, first:

1. Carefully consider your motivations, expectations, costs and resource commitments

2. Go in with yours eyes open. Do your research and learn about potential problems.

3. Simulate the process before committing to it, and…

4. Shop around carefully, just as you would in any other sourcing situation.

CAUTIONS

1. **Avoid "Fads".** Use 3PL services because there is a sound business case, NOT just because competitors do or some "Lean Guru" likes the idea.

2. **Don't abdicate your "core competencies".** Your business will succeed because you can do certain things better, faster or cheaper than any body else in your market. Make certain that none of those things are on a 3PL's contract list.

3. **Don't glorify what is simply an "overflow warehousing" service.** If all you need is an overflow warehouse, outsourcing the need makes location control and accessibility both harder and slower. You also can fall victim to the "Out of sight, out of mind" trap, and forget to shrink the root-causes of excess/surplus. You are better off investing in supplier lead time, minimum order mandates, packaging and delivery-reliability improvements.

4. **Don't expect greater speed**. Working with a 3PL involves extra steps and greater movement distances.

5. **Don't expect greater flexibility**. Like ANY supplier, 3PLs have a lead time for doing what they do. Any changes during or just prior to that interval lead to chaos, just as they do in your own operation… maybe even more!

6. **Don't expect lower costs**. A 3PL may be able to do certain things more efficiently than you, but they also plan to make a profit, and they will mark-up those costs to "market value" levels (as in a quality seafood restaurant).

7. **Don't share**. Trying to work effectively with a 3PL that services several clients from one facility seldom succeeds. Distances are greater than necessary, location and labeling systems are of their choosing, inventories can be mixed, cycle-counting gets overly complicated, team loyalties erode, and response lead times get longer.

8. **Fully explore I/T and communications alignment/technology needs.** I know one client who contracted with a 3PL, then later discovered that their computer systems were nearly incompatible. Their system could send data to the 3PL's, but reverse transfers were blocked, and neither partner could query or correct the other's data files. They also assumed that simple cell phone services would do as well as the recommended PTT system (push-to-talk closed network). Corrective programming and systems retrofit cost a *FORTUNE*.

9. **Explore item labeling needs and systems.** Even if you both have bar-code systems, are they aligned? I know of 4-5 types of readers, 3-4 printer types, and at least a dozen code formats (some numeric-only, some full "Qwerty", some "open" for distance reading, and some "hi-density" to maximize info/space). What formats are supplier-coded labels in now? What formats are least expensive if you are trying to increase supplier bar-coding?

10. **Don't try to solve problems by outsourcing them**... especially supply chain issues like lead time or delivery reliability. If they are issues, adding another player in the process will just add complexity and confound resolution.

SIMULATE

Before outsourcing any process or task, make certain you fully understand it, and can successfully do it yourself (even though with questionable efficiency).

You can form a team to spend months brainstorming and debating all this, or you can learn by simulation... Create a temporary, in-house team to perform the intended 3PL functions, and require them to operate as if they were an independent service agent, including separate personnel, I/T and communications systems.

Pay special attention to communications, data-quality, and response-time problems as they arise. If they bother you, they will drive a 3PL insane.

SHOP CAREFULLY

1. **Beware of novices and "want-to-be's".** Public warehouses have been around practically forever, and many aspire to 3PL status (and revenue potentials). Don't provide them with a learning ground and "mistake-based" education. Hire trained, experienced, reputable professionals.

2. **Beware of the lowest bidder.** Avoid contracting with the first 3PL you encounter, and take the time to meet several. Good ones know their worth and will be competitive, but quotes that seem too good to be true usually are.

3. **Favor nearby operations, or operations willing to locate nearby to serve you.** Communications and response time are at issue here, especially in emergencies or when a component or item is damaged or defective and needs fast replacement.

4. **Favor 3PLs with their own fleet of shuttle/transfer vehicles.** If you are holding them accountable to perform, they need to have control of key elements in the process. Using yet another contract service for transport guarantees confusion and disappointment.

5. **Require and check references.** Especially investigate factors such as ease of doing business, accounting & documentation accuracy, assigned team continuity or turnover rates, safety/damage/security performances, insurance coverage, etc.

Enabling (or blocking) Success

I'm an Engineer, but I'm also sensitive to what kept me working for the same firm, and on similar challenges for over 40 years. During that time, I was also fortunate to do a "rotation" with the Global Operations Development Team within Corporate Human Resources, where human performance support systems were a serious area of study.

Unfortunately, Inventory or Materials Managers and/or Materials Buyer/Planners are too often like the comedian Rodney Dangerfield, who is best remembered for his lament that "he never got any respect". Perhaps this is due to organizations being biased in favor of people who actually produce product, and their failure to recognize that materials planning and management could easily fill a 6-year course of University studies. As a consequence, materials planning, procurement and inventory control teams exhibit lots of frustration, high turnover, and mediocre results.

Operations that have succeeded at becoming "lean", and at business in general, place as much importance on Materials & Inventory Management as on any other area of expertise. In fact, the most senior Inventory Planning and Value-Chain Management positions are sometimes reserved as a prestigious reward and honor for their most accomplished people.

I have observed, and concluded, that there are at least 8 critical components to recruiting, developing and retaining World-Class M&I Management talent:

1. **Significant span of control, position status and a clear career path.** M&I Managers need to have peer status with Finance, Production and Operations Managers in order to accomplish anything worthwhile. Similarly, they either need to have a real and equal opportunity for promotion, or the position needs to have "reward-assignment" status.

2. **Adequate staff, tools and improvement budgets.** This includes I/T systems, software (e.g. IQR®), A-B-C classification tools, Cycle-counting teams and processes, Bar-Coding and Labeling systems, storage racks and bins, purpose-designed carts, communications systems. Most importantly, General Management and Finance must abandon the notion that I&M management personnel are "indirect headcount" and prime candidates for any needed "contractions". If anything, the OPPOSITE attitude is needed... Most operations can flex and work wonders *IF* they have needed materials and close control of on-hand inventories!

3. **Formal Job Descriptions, responsibilities and team memberships (e.g. S&OP).** Ambiguities or "turf wars" do more to kill organizational teamwork than just about any other issue. Toyota even encourages management teams to pre-resolve how to manage the organization chart's "white spaces".

4. **Appropriate, formal training and benchmarking opportunities.** M&I staff are NOT born knowing how to operate MRP/DRP/ERP/WMS computer

systems. They need training by the *vendor*. Likewise, they are unlikely to learn new methods from in-house experts. They need to belong to organizations such as APICS, and to have a budget to attend professional conferences, clinics, and benchmarking opportunities... Just like Sales, Engineering, and the other professional groups!

5. **Clear accountabilities and performance recognition.** This means competitive compensation, organizational awareness & respect for successes, and motivation + extra recognition for innovation and continuous improvements.

6. **"Lean" and flow-focused performance metrics.** Good examples include Inventory Record Accuracy, ASWO counts and costs, supply-chain lead times, inventory turns and average days of supply. Poor choices include any metric that assumes or invites capacity production without regard for demand levels... metrics such as burden absorption, standard labor cost, machine utilization, unit output per worker, etc. These should be subordinated or abandoned. Most certainly, metrics like direct/indirect labor ratios have NO legitimacy, and should be banned! Remember, people focus (and often manipulate) what gets measured, and the Law of Unintended Consequences almost always applies. Remember also, that the same good/priority metrics need to apply across the entire organization. Differing standards lead to conflicting goals and strategies.

7. **Stay the course!** "Lean" attitudes, improvement programs and processes CREATE flexibility, ENHANCE responsiveness, REDUCE lead times and CONTAIN costs... which are precisely what is needed when external pressures mount. Commit to NOT making M&I, training, quality and continuous improvement personnel the 1st targets of any necessary "force reductions". "Indirect" does NOT mean non-value-adding or non-critical.

8. **False Assumptions re Disappointing Results.** People do NOT get up in the morning plotting how they can frustrate the organization during the day. People inherently want to do a good job if for no other reason than self-respect. IF there is a problem, the odds of a process-based root cause are at least 85% and the odds of personal failures less than 15%. Even then, much of the 15% can be explained by inadequate training.

Some Concluding Thoughts

Managing inventory in a "Lean" environment, eliminating ASWOs and minimizing surpluses, is *NOT* "rocket science". Rather, it relies on committed, everyday "blocking & tackling" by the entire organization... including Management, Forecasters, Buyer/Planners, Finance, Operations and, yes, even HR!

Mostly, it relies on Leadership with Vision about what could-be, and the energy to recruit and excite others. Is that you?

Hopefully, by reading and considering this guidebook, you have found some new ideas for working smarter rather than harder:

- The basics of "lean", including the 10-steps and 7-wastes
- The (familiar?) inventory management process, but with some new inputs
- The importance of clear ownerships and accountabilities
- ASWO root causes from around the World
- Surplus root causes (often the same as ASWO causes)
- Robust location systems design principles
- Probability-based forecasting of *RANGES*
- Flow-focused metrics and the hazards of many traditional measures
- Usage-value-based A-B-C item classification and management systems
- A-B-C based Cycle-Counting systems for on-hand records accuracy
- Working successfully with 3PLs
- Enabling success via both people and process

All of these initiatives have yielded great results in numerous sites, of all types and sizes, all around the World, but no site has (so far as I know) implemented them all. This means you could be the first... *IF* you can avoid or "end-run" the barriers the others encountered, including...

- General Management's seemingly universal need for "instant gratification"...GT^3... Good Things Take Time.
- Restricted resources, especially people. Headcount and direct/indirect labor ratios are utter nonsense, as you likely know, and as confirmed by the University of Michigan's 2002 study of the subject.
- Inappropriate performance metrics, and a focus on things other than flow.
- Outstanding results on first initiatives "took the edge off" implementing the rest, even though none of the initiatives can yield their full potential without the help of all the rest.
- Unfortunately, "change" is too often postponed until organizations "hit the wall" and have no choice, and "staying the course" is equally abandoned once any success provides minimal relief.

- Distractions (like the recent, Global Economic turndown) captured General Management focus. (Even though one of their key issues was reducing inventory!)

If this guidebook helped you, put a copy into the hands of your colleagues, and work with HR to make reading it part of new employee development/orientation.

Remember the principal rules of inventory management defined in the introduction:
Rule 1: **NEVER** stop production line or block a sale for lack of materials.
Rule 2: **ALWAYS** work to minimize material and product inventories.
Rule 3: **AVOID** potentially dysfunctional or counterproductive performance metrics.
Rule 4: **LEARN** from every situation and player (including the occasional idiot).
Rule 5: ~~IF~~ **When** **FRUSTRATED or CONFUSED**, you probably ignored Rule 3.

Some potentially useful Excel® spreadsheets have been illustrated in several places in the guidebook's text. If you would like to have actual samples, instructions for getting them are in the appendix.

This guidebook is by no means exhaustive on potential tools, or on any of the tools presented. Each could fill volumes by themselves. If you have additional questions, suggestions, or would simply like to discuss your own challenges and plans, feel free to contact me by e-mail or phone (full details are on www.ILSDllc.com).

My sincere thanks to you for taking the time, and having the interest, to read "ASWO!", and please accept my sincere best wishes for success as you design and deploy your improvement plan.

It was Albert Einstein who properly quipped, "Insanity is continuing to do what has been repeatedly done before, and expecting a different outcome."

Jeve Morman added, "Status quo is Latin for the mess we're in."

Ernest Hemingway added, "Never mistake motion for action."

Oscar Wild observed, "I am not yet young enough to know everything."

Finally, Gen. George Patton advised, "If everyone is thinking alike, somebody isn't thinking.

Eric Matson

APPENDIX

Inventory Quality Ratio (IQR)® Management Software

On several occasions, I have mentioned IQR® Software as a tool most worthy of purchase and use. I am not a stockholder of IQR® International, but I am an excited user of many years, and hope to save you considerable search time.

IQR® is written and offered by IQR® International in California, and is available from them in two versions: One for stand-alone installations, and another for multi-user or multi-site access to a shared database.

A single-site, 10-user site license has an initial cost of just over US$30K (less than an additional buyer/Planner), and a modest annual maintenance fee.

What is important is the amazing time and cost savings that come from use.

Where I have installed it and trained users, the most popular features have been:
- Tracing of Inventory values, by category, over time (levels and trends)
- Automatic assignment of usage-value-based A-B-C (+ D-E-F-G-H-K) management classes to each item, AND the calculation of by-item maximum-quantity goals for targeted turns.
- Fast, accurate calculation of scrap, obsolescence and shrinkage financial reserves needs
- Ability to "shuffle the deck" and review values by class, category, material type, Buyer/Planner, etc.
- Ability to write and execute custom, by-part or item reports, and re-run these reports at any time *without* being linked to the MRP/DRP/WMS main system.
- Ability to see ALL available information on an item on a *single screen*... no flipping among multiple screens to get a complete picture!
- Ability to get overall or by-item information about...
  o Potential shortages
  o Apparent surpluses
  o Non-moving or slow moving items
  o Dysfunctional system data-points such as...
    - Items with zero days availability lead time
    - Items with apparently negative on-hand quantities
    - Items with apparently negative requirements or usage
    - Items with missing source, Buyer or Planner codes
    - Items with hurtful minimum order quantities
    - Items with hurtful safety stock settings
    - Etc... imagination is the limit!

More complete information is available directly from IQR® International at www.IQR.com. If you end up talking to Gary or Brian, tell them Eric sent you!

Here's some exhibits I use during user training to "wet your appetite":

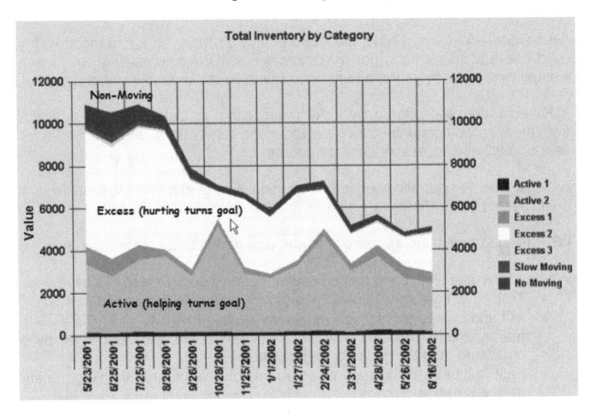

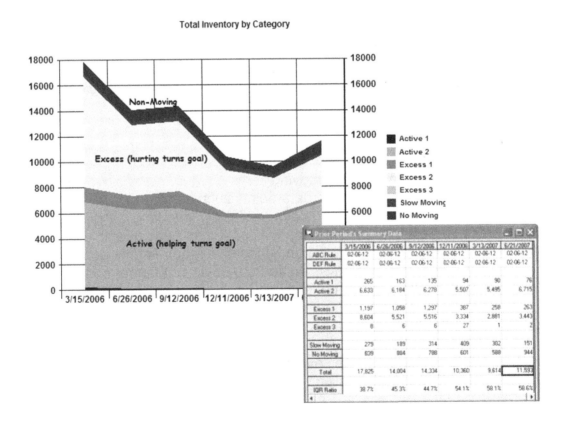

Reserve requirements calculations in 2 seconds? (help the Finance Dept at plans time!)

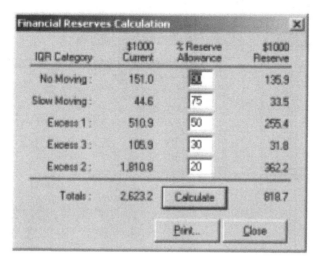

Financial Reserves Calculation

IQR Category	$1000 Current	% Reserve Allowance	$1000 Reserve
No Moving :	151.0	90	135.9
Slow Moving :	44.6	75	33.5
Excess 1 :	510.9	50	255.4
Excess 3 :	105.9	30	31.8
Excess 2 :	1,810.8	20	362.2
Totals :	2,623.2	Calculate	818.7

Print...   Close

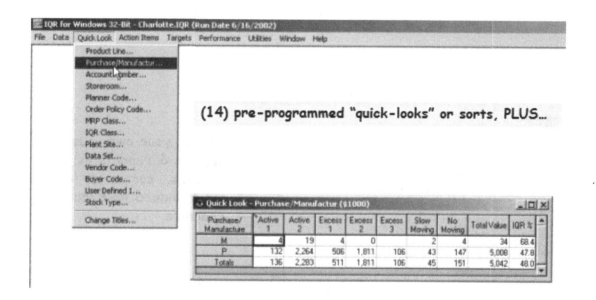

IQR for Windows 32-Bit - Charlotte.IQR (Run Date 6/16/2002)

File   Data   Quick Look   Action Items   Targets   Performance   Utilities   Window   Help

Product Line...
Purchase/Manufactur...
Account Number...
Storeroom...
Planner Code...
Order Policy Code...
MRP Class...
IQR Class...
Plant Site...
Data Set...
Vendor Code...
Buyer Code...
User Defined 1...
Stock Type...

Change Titles...

(14) pre-programmed "quick-looks" or sorts, PLUS...

Quick Look - Purchase/Manufactur ($1000)

Purchase/ Manufacture	*Active 1	Active 2	Excess 1	Excess 2	Excess 3	Slow Moving	No Moving	Total Value	IQR %
M	4	19	4	0		2	4	34	68.4
P	132	2,264	506	1,811	106	43	147	5,008	47.8
Totals	136	2,283	511	1,811	106	45	151	5,042	48.0

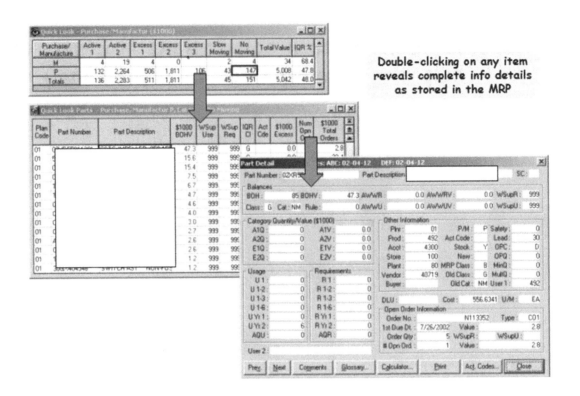

Double-clicking on any item reveals complete info details as stored in the MRP

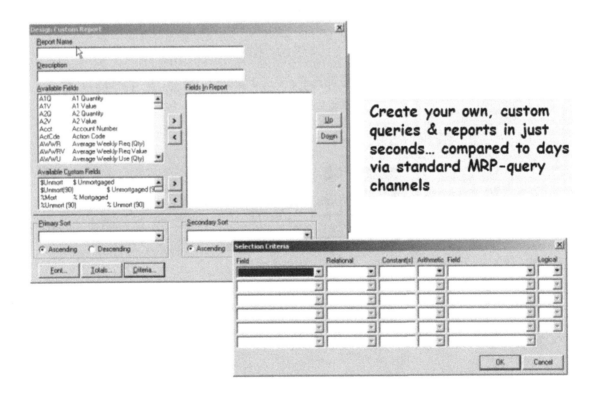

Create your own, custom queries & reports in just seconds... compared to days via standard MRP-query channels

Name	Description	Date Revised
"Warm List"	AAA A Shortage (Needs Date input: today+14)	08/15/06
Zero Lead Times	AAA B Shortage	06/22/05
Overdue Deliveries?	AAA C Shortage (Needs Date Input: today-4)	08/15/06
Slow & Non-moving Parts	AAA D Dormant	09/28/05
Open PO's for Non-moving Items	AAA E Dormant (needs date update in criteria)	08/15/06
Premature Deliveries?	AAA F Excess (needs date inputs)	08/15/06
Excessive Min. Buys	AAA G Excess	06/22/05
Excessive Safety Stocks	AAA H Excess	06/22/05
Long Lead Times	AAA I Excess	03/11/05
Needed but > 30 Days Supply	AAA J Excess	08/16/06
Used but > 30 Days Supply	AAA K Excess (ongoing use assumed)	09/29/05
Zero Unit Cost	AAA L Integrity	07/01/05
No Vendor Code	AAA M Integrity	06/22/05
Under-the-Radar at ETP	AAA N Integrity	06/30/05
P/M Code Mismatch	AAA O Integrity	03/11/05
No Buyer Code	AAA P Integrity	06/22/05
No Planner Code	AAA P2 Integrity	04/11/06
No User-1 Code	AAA P3 Integrity	04/11/06
P/N's and Buy $ by Buyer	AAA Q Integrity	03/11/05
Compressors	AAA R	07/27/05
Slow & Non-moving FG Units	BBB A Dormant	03/10/05
Slow & Non-moving FG Access	BBB B Dormant	07/27/05
FG Unit On-Hand Values	BBB G Overall	03/10/05
Unmortgaged FG Unit Values	BBB H Overall	06/29/05
FG Access. On-Hand Values	BBB I Overall	03/10/05
Unmortgaged FG Access Values	BBB J Overall	03/10/05
FG Unit Penny Stocks	BBB K Integrity	03/10/05
FG Access. Penny Stocks	BBB L Integrity	03/10/05

...or choose from over 30 already developed by user experts

<u>Report Categories include</u>:

- Potential shortages
- Dormant SKU's and Materials
- Excess/Surplus holdings
- Problem Purchasing parameters
- Database errors & omissions
- (OK for WIP, Parts, FG)

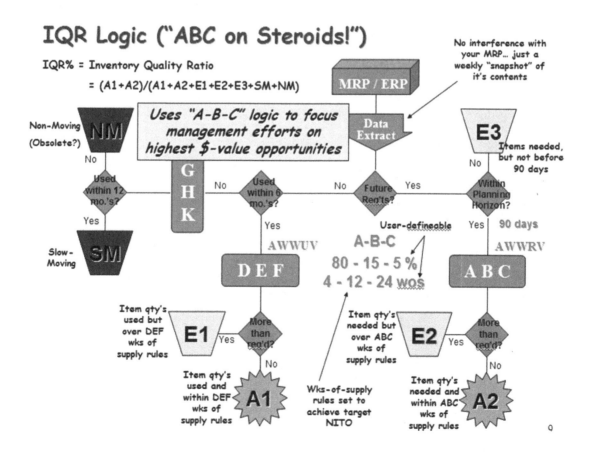

133

Requesting Sample Copies of Excel® Spreadsheets Illustrated in "ASWO!"

Excel® Spreadsheets described in "ASWO!" include:
- The Inventory Management Process
- Forecast Accuracy Tracker
- Days-of-SupplyTracker
- ABC Classification Analysis
- ABC-based Wks-of-Supply calculator
- Shortage Analysis Spreadsheet
- MRP/DRP Data De-bugger

I am more than happy to share actual samples of these with "ASWO" readers, at no cost, *IF* you will send me either an e-mail address (for electronic file transfer), or an (inexpensive) USB flash drive and self-addressed, pre-stamped return envelope (all the files occupy under 4 Mb).

My current e-mail address is: ematson@twcny.rr.com. If that changes for any reason, the new address will be listed on our consultancy's web page, www.ILSDllc.com

Please feel free to "borrow shamelessly" and/or customize these, *but* any and all credits for the source and "ASWO!" guidebook will be greatly appreciated.

Glossary of Terms and Acronyms

ASWO!©	"Ah, Shucks, We're Out!" - Expression of dismay about a shortage
3PL	Third-Party Logistics service provider
5S	Japanese term which translates to: Sort, Store, Shine, Standardise, Sustain
AMED	"Any Model Every Day"
APICS®	Association for Operations Management
Back-flush	MRP computer process for converting completed products to consumed components
BOH or QOH	Balance-On-Hand or Quantity-On-Hand
BPM	Business Process Map
CC	Cycle-Count
Concurrent Engineering	Cross-functional teaming
COPIRA	Cost Of Poor Inventory Accuracy
CRSD	Customer Requested Shipping Date
CTQ	Critical To Quality
Cycle-Counting	Scheduled, usage-value-biased counting of location contents
DRP	Distribution Requirements Planning System
EOQ	Economic Order Quantity (usually mis-calculated)
ERP	Enterprise Requirements Planning System
FAST!©	"Forecasting And Scheduling Tips" - A Schmidt Creek Paddler's Guidebook
Faurie	Noth Dakota bass-killer, a.k.a. "Cap'n Hook"
FG	Finished Goods
FIFO	First In, First Out
FSO	Field Shipping Order (inter-site transfer authorization)
F-Word	Forecasting?
GM	Gross Margin
Heijunka	Load, Schedule or Task Leveling (Japanese)
I & M	Inventory & Materials
IBF®	Institute of Business Forecasting
IE	Industrial Engineer
ILSD llc®	Integrated Lean Systems Deployment llc - Affordable Lean Consulting firm
IPOSA	"In Process Off-Standard Authorization" (BOM item waiver)
IRA	Inventory Record Accuracy
JIC	"Just-In-Case" (typical)
JIT	"Just-In-Time" (desired)
Kanban	"Signal" (Japanese), usually denoting a standardized, visual replenishment system
Kentucky Windage	Estimate of measurement error and correction factor (from rifle shooting)
KISS!©	"Keep It Super Simple!" - Schmidt Creek Paddler's Guidebook
LE	Latest Estimate

Lean	Meaning streamlined, flowing and waste-free.
Line Balancing	Distributing process tasks so that each contributor needs about the same time
Logistics	The systems and processes used to locate, contain and move materials and inventory
M&I	Materials and Inventory
Min / Max	An item inventory control system with minimum and maximum on-hand quantity boundaries
MOQ	Minimum Order Quantity
Mortgaged Inventory	Inventory with matching needs in the schedule
MPS	Master Production Schedule
MRP	Manufacturing Resources Planning system
Muda	Waste (Japanese)
NIL	Not In Location
NITO	Net Inventory Turn-Over
NM	Non-Moving
OMQ	Order Multiple Quantity (e.g. must order by the dozen, hundred-weight or case)
Outsourcing	Placing an internal item or task with an independent Supplier
PBOM	Planning Bill Of Materials
PFEP	Plan For Every Part
QCPC	Quality Control Process Clinic (or Chart)
Rate-Based Planning	Flexible but bounded scheduling and control linked to customer demand variabilities
RFID	Radio Frequency Identification (system, tags, etc)
RRCA	Relentless Root Cause Analysis
Schmidt-Creek Paddler's Guidebooks©	A series of guides designed to explain ignored lean topics
Sensei	Guide… Teacher… One who has made the journey before (Japanese)
SKU	"Stock-Keeping-Unit"
SM	Slow-Moving
SOP or S&OP	Sale and Operations Planning… team or meeting
Stochastic	Based on or involving statistical analysis or inference
Unmortgaged Inventory	Inventory without matching schedule needs or requirements
VMI	Vendor-Managed Inventory (but usually on site)
VSM	Value-Stream Map
WIP	Work In Process - typically unfinished product and components
WMS	Warehouse Management System